PLEASE STAMP DATE DUE, BOTH BELOW AND ON CARD

DATE DUE	DATE DUE	DATE DUE	DATE DUE
4.20.90			
JUN 16 1990			

GL-15(3/71)

Introduction to
THE THEORY OF
SIMILARITY

Introduction to
THE THEORY OF
SIMILARITY

A. A. Gukhman

Translated by SCRIPTA TECHNICA, INC.

TRANSLATION EDITOR

ROBERT D. CESS
DEPARTMENT OF THERMAL SCIENCES AND FLUID MECHANICS
STATE UNIVERSITY OF NEW YORK
STONY BROOK, NEW YORK

1965

ACADEMIC PRESS New York and London

ACADEMIC PRESS INC.

111 FIFTH AVENUE

NEW YORK, NEW YORK 10003

United Kingdom Edition

Published by

ACADEMIC PRESS INC. (LONDON) LTD.

BERKELEY SQUARE HOUSE, LONDON W. 1

Library of Congress Catalog Card Number: 65-24999

Originally published as
"Vvedenıye v Teoriyu Podobiya"

by State Press "Vysshaya Shkola" (Higher Education), Moscow, U.S.S.R., 1963.

PRINTED IN THE UNITED STATES OF AMERICA

TRANSLATION EDITOR'S FOREWORD

Professor Gukhman's book gives a very thorough presentation of similarity theory with particular emphasis upon physical interpretations. The basic ideas are applied to several illustrative problems, and these serve to clearly emphasize the techniques involved in the use of generalized analysis.

The terminology employed in the book is quite compatible with that common to the English-speaking reader. Although all equations and figures have been reproduced from the Soviet edition, this in turn presents little difficulty since the nomenclature which Professor Gukhman has chosen is quite universal.

A subject index, which is characteristically absent in Soviet books, has been added to the present English edition.

R. D. Cess

Stony Brook, New York, May 1965

FOREWORD TO THE ENGLISH EDITION

I am delighted that an English translation of my book is being published. To a considerable degree the theory of similarity owes its creation and development to a series of publications whose authors include such giants as Newton, Rayleigh, Buckingham, Bridgman, and other prominent scientists who wrote in English.

In this book I have attempted mainly to indicate the essential features of the theory of similarity as a special method of quantitative investigation. My desire has been to let the reader feel the spirit of this science, to assist him in understanding the physical foundations of its main ideas and principles, and in grasping the part it plays in the modern system of exact science. In addition, I have attempted to develop a form of presentation which makes it possible to concentrate on the physical content of the theory without losing logical completeness and exactness.

It is not for me to judge how well these objectives have been realized. In any case, I will be pleased if the book now presented to the reader proves useful to him and stimulates his interest.

A. A. Gukhman

FOREWORD

This book is devoted to the fundamental concepts of the theory of similarity, which can be regarded as the science of characteristic generalized variables for any given process. The methods of deriving and applying these variables are considered. This view of the nature of the theory of similarity extends throughout the book and determines the general plan of the presentation and the choice of material. Considerable attention is paid to the fundamental ideas and their interpretation.

The concept of the theory of similarity is extremely simple. It can easily be mastered, and the technical habit of using it is readily acquired. However, even long experience in the formal application of the theory of similarity cannot teach its most important and less obvious applications and abolish routine and even outright errors. It is only by a study of the physical ideas which form the basis of the theory of similarity and which are incorporated in its method of use that it is possible to arrive at the correct view of the theory of similarity as a tool of quantitative research and of the true possibilities and rational applications of its methods. In order to become proficient in the use of similarity theory it is necessary to understand its physical substance first.

I have attempted to write this book in such a way as to make clear the very close relationship between the fundamental physical concepts of the theory of similarity and its mathematical implications, so that the theory of similarity is perceived as a system of ideas having a clear physical significance. The whole layout of the book has been selected with this in view. The various considerations required to provide the theory of similarity with the desired formal exactitude and completeness are considered as fully as necessary. However, these are preceded by a statement of the general theory and its concrete applications, so that they are presented on the basis of a well-developed view of the physical aspects of the problem as a whole. Dimensional analysis is presented in this way. The theories of similarity and dimensional analysis are considered as a unified entity but different modifications of the same procedure of investigation, whose differences are due only to the different amounts of preliminary information available about the process being investigated.

It is necessary to explain the layout of the material, particularly with respect to the amount of space devoted to concrete applications of the theory. The book was planned (as indicated by the title) as an introduction to the theory of similarity, that is, as a small book intended to acquaint the reader with the fundamentals of the theory and with the techniques of using it. An investigation of various processes was not an objective in itself. Specific applications can only be used here as a means of particularizing the general arguments and, in some cases, as a means of arriving at important new results by studying special problems. However, it would be wrong to reduce the applications to a series of unconnected illustrations. Instead, it is much more useful to consider in systematic form a large and fairly complete problem. Under these conditions the stage is set for a serious analysis of the effect which the theory of similarity can have on an investigation as a whole, from establishing the problem to its final solution.

Thus, we will consider only a very limited circle of problems. On the other hand, it is of the greatest interest to investigate other processes which are important theoretically or through their applications, and, naturally, as many of them as possible have been included, though only the most characteristic problems have been considered in order to avoid making the book too long. The various applications of the theory of similarity are important in practice and interesting from a theoretical point of view; they are so numerous that they could provide the subject matter for a whole series of books. The applications of the theory of similarity in the study of transfer processes in flowing media deserve special mention, and I propose to make these the subject of another book which will form a continuation of the "Introduction to the Theory of Similarity."

In conclusion, I would like to point out that the present book represents an extension of the ideas given by the booklet "The Substance of the Theory of Similarity," which was written by E. A. Ermakova and myself for the students of the Moscow Chemical Machine Construction Institute (Institut Khimicheskogo Mashinostroeniya).

A. A. Gukhman

CONTENTS

INTRODUCTION

Quantitative investigations play a great part in the scientific and technical activities of scientists. Quantitative predictions are playing an increasing part in every field of knowledge, and there is a continuous process of transition from qualitative statements towards quantitative laws. Modern scientific thought is based on the conviction that theoretical concepts have a real and exact nature only when they are expressed in the form of quantitative relationships. It is only when this is done that it is possible to use theories effectively for practical purposes. As a result, the problems of the means available for quantitative investigation, and of the degree of agreement between these and the actual content of the problems arising in the development of science and its technical applications, are of great importance. Thinking about these problems will show that there is an important gap between the possibilities of modern mathematical techniques and the requirements which arise in practice. This gap has produced a very difficult situation.

A sufficiency of quantitative information for any theoretical investigation or technical application is possible only if each of the quantities defined as variables for the process being investigated is known when the problem is set up. This makes it possible not only to calculate the values of all the variables for the actual conditions of the problem, but also to deduce the effects which changes of these variables will cause. In such cases, we say that the quantitative laws giving the relationships between the variables have been established. The investigation can be completed at this point if it is possible to give a complete analytical description of the problem.

In the vast majority of cases attempts to obtain analytical solutions of problems meet with considerable difficulties. Usually it is possible to carry an analytical investigation to a conclusion only at the cost of making important simplifications either in setting up the problem or in the course of the solution. The results obtained are therefore approximate estimates at best, and at worst they may be incorrect in important respects and be the source of serious errors. It must be admitted that at present purely analytical investigations are possible only in principle, and cannot be realized in practice because of the complexities of the problems and the stringent requirements of accuracy and detail in the solutions.

Thus the natural path to the establishment of general quantitative relationships appears to be blocked. This does not mean that in general it is impossible to obtain quantitative results. On the contrary, the quantitative side of the study of problems is receiving more attention than ever before, though the nature of the investigations has changed considerably. *Numerical* methods of solution are of very great importance. Remarkable advances have been made in this field, and at present it is possible to give a numerical solution to many problems, regardless of complexity, to any required degree of accuracy.

However, analytical and numerical solutions are by no means equally valuable. The series of numbers which are obtained as the result of a numerical solution express a great deal of very valuable information which can be used successfully. However, they do not indicate the interrelationships which characterize the problem being investigated. Of course, an analysis of the numerical results always makes it possible to find some of the relationships. It is possible to set up equations which approximate these data to some degree of accuracy, but it is obvious that such equations are no more than empirical correlations. The isolated partial relationships give the connection between individual variables but not a general, overall equation; they make it difficult to see the problem as a whole and do not lead to a complete and clear picture of the process. Obviously, they become less clear and of less practical value as the number of variables in the problem increases. In principle, nothing is changed if we study the results of an experimental investigation instead of the numerical solution of the problem. The only difference is that the values of the variables linked by the empirical relationships are obtained by measurement rather than calculation.

Thus, numerical methods (or direct experiments) are insufficient for the determination of general laws. At best, when these calculations or experiments are used it may be possible to arrive at a series of partial relationships between the separate variables. It is generally impossible to establish the system of relationships expressing the fundamental quantitative laws of the problem which are adequate for the important and general physical ideas laid down in formulating the problem. Fortunately these methods can be made considerably more powerful by using other methods of investigation which are based directly on an analysis of the physical mechanism of the process being studied; these lead to important relationships which cannot be derived in any other way. A combination of these relationships and the results of numerical solutions or experiments is very fruitful and lends greater depth to the whole problem. We will attempt to explain the main features of this line of attack.

It is clear from the remarks above that the number of variables which exist for a process being investigated is of the very greatest importance. The necessity of dealing with a large number of different quantities does not present particular difficulties if the solution can be carried to a conclusion in analytical form. In this case the quantitative laws of the process are clearly expressed, and the effects of all the variables can be found directly. On the other hand, if methods are used which lead to specific numerical relationships instead of analytical expressions, the investigation becomes increasingly complicated as the number of variables increases. When the number of variables is large it is very difficult in practice (or even quite impossible) to reduce the results of the solution to a systematic form, to find the hidden relationships, and to combine these relationships into general quantitative laws. The problems that are typical at present include many that are of very complicated physical content. When such problems are studied it is necessary to introduce a multitude of different quantities, each of which is regarded as an independent variable. Here we must look into the reason for the fundamental difficulties arising in quantitative investigations.

Among these concepts we must introduce a new idea which changes the formulation of the problem and gives the investigation a trend which is of interest to us from the beginning. On the basis of very general physical arguments it can be shown that the multiplicity of relationships is not a natural property of the physical problem being investigated. It appears that actually the effects of the separate factors represented by the different quantities do not come about separately but in groups, and that in fact it is necessary to deal not with the separate quantities, but with definite combinations of them, depending on the process being studied. It is possible to give a method for deriving such groups—a method which enables one to derive the relationships between the various quantities and to combine these into well-defined groups on the basis of a direct analysis of the problem. Such groups have clearly defined physical significances, and they represent a special type of variable which is characteristic of the particular problem.

There are important advantages in converting the ordinary physical quantities into groups. These groups should be well defined and should depend on the nature of the process. First of all, there is a reduction in the number of variables. In addition, when a problem is investigated in terms of these quantities, which represent the effects of groups of factors rather than those of individual factors, it is easier to find the interrelationships characterizing the process, and the whole quantitative picture becomes clearer. Finally, the new variables have other important features also. It is obvious that a given value of a group can be obtained as a result of innumerable combinations of the quantities of which it is comprised. As a result, a given value of the new variable corresponds not to just one combination of the initial quantities but to an array of such combinations. This means that in considering a problem in terms of the new variables, we are not studying just a single special case, but an infinite number of different cases related by some general property. Thus the new variables are essentially general. As they are varied, the whole analysis acquires a generalized nature.

The replacement of ordinary by generalized variables is a fundamental feature of the procedure which we will consider. This procedure is known as the theory of similarity and dimensional analysis. Both these terms indicate the technical methods on which the system is based (and, to some extent, the origin of the methods), rather than its real essence. It would be more correct to call it the *method of generalized variables*. This method is the subject of the present book.

Chapter I

FORMULATION OF THE PROBLEM

1. THE PHYSICAL MODEL OF A PROCESS. THE FUNDAMENTAL EQUATIONS OF THE PROBLEM. CONDITIONS FOR UNIQUENESS OF SOLUTION

A study of the general quantitative laws of a phenomenon can only lead to useful results if it is based on sufficient physical information. The quantitative investigation is preceded by the long and complicated process of formulating the physical concepts. This process is often closely allied to experimental work. At some stage in its development it leads to a model of the phenomenon, i.e., to a physical scheme which postulates the applicability of some quantitative physical law either directly, or with special additional hypotheses. A model of a phenomenon represents a schematization of the true occurrences. The degree to which this is the case depends on the general design and purpose of the analysis and on the completeness and accuracy required in the solution. In every case the important features of the investigation must be specified. It is essential that the final model should present the most important features of the phenomenon; the model does not have to reproduce all the minor details. Thus a quantitative analysis always deals not with a real phenomenon, with all its specific complexities, but with the results of a more or less complete schematization. Nevertheless, there are considerable difficulties in passing from a physical model to its quantitative description; these are caused by the very nature of the problems which are studied. There appear to be two possibilities here.

Sometimes the study of the physical mechanism of a process leads to clear and specific concepts from which it is possible to derive equations which correspond exactly to the model which has been assumed. In this case the investigation is based on equations which define the process and are known when the problem is formulated.

However, the degree of clearness of the physical representation required for deriving these equations cannot always be attained. Under these conditions one has to make do with relationships which characterize the process in some general way only, and which, as a result, convey less information about its physical mechanism. The method of generalized variables can be used successfully in both cases, regardless of whether it is necessary to start from the equations for the process itself or from relationships of a more general nature. If we are dealing with the application of the method to an investigation which is of immediate interest, rather than making a detailed analytical investigation, the difference in the amount of preliminary information available may be quite substantial, but is usually not of great importance. Perhaps it would be best to speak only of certain aspects of the application of the methods. Initially we will base all our arguments on the assumption that the equations describing the process are known, and show how the investigation is developed in this case. The application of the method of generalized variables to cases where equations cannot be set up will be considered in a later chapter.

Thus we will assume that the equations describing the process being studied are known. In most cases they will be differential (or integral, or integrodifferential) equations. These equations, which express the physical representation of the problem, are known as the *fundamental equations* of the problem. Very often the problems to be investigated are so complicated that they cannot be studied in sufficient detail on the basis of only one physical law, and it becomes necessary to consider various aspects of the model and to introduce different physical laws. In such cases, the process as a whole is described not by a single equation but by a system of fundamental equations.

An important feature of the present problem is that it is possible to investigate phenomena which are the result of a large number of diverse factors, each of which can be expressed in the equations. In this way equations are obtained which contain a large number of quantities of various physical types.

Thus a typical equation in which we are interested may have a complex analytical structure (differential, integral, or integro-differential) and may include a large number of different types of quantities. On the other hand, the physical laws which are the sources of the fundamental equations of the problems to be studied are very simple both in concept and in form. Naturally, since each of these equations represents a special form of the corresponding law which is applicable to the model which is assumed, each must express some simple, readily understandable quantitative relationship. Actually, in this sense any of the equations has a clear and distinct meaning. For example, one of the fundamental equations of fluid mechanics—the continuity equation—is basically a law of conservation of mass (applied to a moving medium). The fundamental equation of the theory of heat conduction is simply a special case of the law of conservation of energy, and can be reduced to the statement that the change in the internal energy of an element of a body is equal to the amount of heat exchanged with the rest of the body. The dynamic equations of motion (or the Navier-Stokes equations) are a specialized form of d'Alembert's principle, and therefore express the very simple condition for the equilibrium of certain combinations of forces. We will have to be content with these few examples. Obviously it would be possible to give the significance of any other equation in this way, no matter how complicated it might be.

We must stress that the equations being considered are simultaneously extremely complicated and very simple. Of course there is no real contradiction here. The fact is that the properties of the same process are given in terms of quantities of different categories. In formulating the physical laws or explaining the physical meaning of an equation, we return to the specific concepts directly related to the very essence of the problem, such as forces of various types, mass fluxes, energy fluxes, etc. The quantities corresponding to these concepts are the quantitative measures of the different effects of which the processes are composed. Their degree of complexity depends on the mechanism of the phenomenon being studied. However, in all cases these are not considered to be a primary type of quantity in the system of ideas representing the problem being studied; in the course of the investigation, they are expressed in terms of others of a simpler physical nature, and, in the final analysis, the quantitative consideration of the problem is reduced to these. These simple quantities, which can be observed and measured directly, and which in fact we always use in our analytical investigations, experiments and calculations, are distances, intervals of time, velocities, temperatures, physical properties (quantities defining the properties of the medium), and so forth. It is true that tedious calculations or difficult measurements may be required to determine some of these quantities, but they are fundamental to the formulation of the problem and cannot be reduced to other quantities. In this sense they can be termed primary quantities. Both scientists and engineers use these quantities directly, and they must appear in the fundamental equations. The transition from the physical law in its general form to a specific equation in the form of a relationship between the primary variables is possible if a sufficiently detailed model of the phenomenon has been constructed in advance. It is also necessary to set up the fundamental equations of the problem in order to express the general physical laws in terms of the primary variables.

Thus the clear, simple physical idea underlying the general physical law is transformed into an extremely complicated equation because the initial simplicity is inevitably lost on transition to the primary variables in terms of which the fundamental equations of the problem must be expressed. A physically simple relationship between the primary variables can only be established in the most elementary cases. The circumstances are much less favorable than this in the typical modern problem, where the physical effects of interest can only be expressed in terms of differential (or even more complicated) equations, and where several such expressions may have to be interrelated to set up the fundamental equations of a problem. The equations obtained in this way contain a certain amount of information as to the relationships between the variables of interest, but this is not in a form useful for drawing direct conclusions about specific interrelationships of the variables.

Thus the difficulties facing us are due to the necessity of expressing the general physical laws in terms of the primary variables. Naturally the question arises as to whether this step is necessary and unavoidable, and whether in general it would be possible to omit the transition to the primary variables and still investigate the problem in terms of variables corresponding to the physical nature of the process which would determine the effects being considered in a more direct manner.

However, such a change would convert the process being investigated into a special, isolated case, artificially separated from other related processes. Besides, under these conditions it would be impossible to study as a whole the different aspects of the same problem. For example, the concept of forces acting at a point is foreign to problems based on the equation of continuity, and cannot play any part in them. An important aspect of processes related to the laws of conservation and transformation of energy is expressed in the equations of energy redistribution in fluid streams. Some of the terms in these equations represent the action of forces. The equations also contain terms which represent heat fluxes which cannot be expressed in any way in terms of forces. The equations expressing various aspects of the complex phenomena which occur in fluid streams are interrelated, since the same physical quantities occur in them (the quantities which we have termed primary variables). The disconnected relationships between quantities having quite different physical characteristics (forces, energy fluxes, mass fluxes) are unified into a single system of equations only because these quantities are expressed in terms of the same variables.

As a result, use of the primary variables is quite unavoidable, even though we have shown that this greatly complicates the investigation. However, we still do not have a clear enough picture of the real nature of the difficulties, since all of the argument above was confined to a family of problems in which only one fundamental equation was analyzed. We will have to consider the problem as a whole.

The fundamental equation of a problem expresses quantitatively a general representation of the physical mechanism of the process. These equations are formulated on the basis of an analysis of the process as a special case to which the fundamental principles of physics are applied. It is impossible for such general relationships to have the same high precision as is encountered in the ideas of single, specific phenomena. The level of abstraction with which it is possible to describe the process when the fundamental laws of physics are used requires a considerable removal from the individual peculiarities of the particular investigation. The scheme used as the model of the process should include only the most important features inherent in the innumerable related but different phenomena. A physical model should refer, both as regards the quantity and the nature of the information which it expresses, to a whole class of phenomena rather than to a single phenomenon. The differences between the phenomena of a given class consist of features which are lost in the process of induction from the experimental facts to the main scheme; the common features are the important aspects defined by the fundamental equations.

Thus the fundamental equations define only those properties of a phenomenon which can be regarded as general for the whole class. These equations cannot contain any information about the specific features of a given phenomenon which distinguish it from the many other phenomena of the same class. The physical causes producing the characteristic differences between the phenomena must be sought in the circumstances of the development of the process, such as the geometrical and physical properties of the systems, the conditions at the moment when the process commences, the effects of the surrounding medium, etc. These must be given in some form which is independent of the fundamental equations and in addition to them. The absence of this information inevitably leads to some degree of uncertainty, corresponding to differences between the general mechanism of the process and the actual form in which it occurs. Depending on the physical nature of the process, the additional conditions necessary for distinguishing a single phenomenon may vary widely in the amount and type of information which they contain. However, from a mathematical point of view, the uncertainty included in the fundamental equations (when they are considered without any additional information) is always uniformly great in all cases.

Actually, it is quite obvious that a valid quantitative description of any phenomenon must satisfy the following basic requirement: for any given values of the quantities which are regarded as independent variables in setting up the problem, there must be only one possible value of each of the unknown quantities (i.e., of the quantities representing the dependent variables). This strictly single-valued relationship between the values of the various quantities leads to the complete determinacy which is characteristic of real phenomena. If an investigation is carried out on the basis of the equations of the problem alone (without using additional conditions), it is impossible to arrive at single-valued relationships. The solution of the equation will be given by some analytical expression—the general integral of the equation—in which the unknown quantity is given in terms of the independent variables. Naturally, all functions which are solutions of the equation must reduce it to an identity on substitution, and this is the only requirement they must satisfy. However, in all cases an infinite number of such functions can be derived, and as a result each of them can be regarded as one of the solutions of the equation. The general integral includes all these solutions, which can be obtained from it as special cases (particular solutions). This characteristic multivaluedness is clearly indicated by the structure of the general integral, since this contains a constant (the integration constant) whose value can be expressed quite arbitrarily. The general integral is converted into one of the particular solutions when the integration constant is given some fixed value. Thus an equation can have an infinite number of different solutions, but only one of these expresses the relationship between the variables which corresponds to a given real phenomenon; this special solution represents a solution of the problem as well as a solution of the equation.

The uncertainty arising in the investigation of equations which are not accompanied by additional information is due to fundamental physical causes. Among these is the fact that the process is not completely defined by the fundamental equations of the problem. Since the initial equations define only the properties which are shared by all the phenomena of the given class, the solutions of these equations must refer equally well to any of the phenomena. As a result the physical significance of the multivalued results is that the general integral of an equation corresponds to a whole class of phenomena rather than to just one. The additional conditions which identify a particular phenomenon within its class must provide the basis for passing from the infinite number of possible solutions of the problem (possible in the sense that each of them can be a solution of the problem since each is a solution of the equations) to the unique real solution. The additional conditions must therefore satisfy the following obvious condition: any of the functions which are solutions of the equations must give a completely defined and unique value of the unknown variable for specified values of the independent variables once the additional conditions are used. This requirement remains valid when different functions which satisfy the equations are obtained, so that the various particular solutions differ from one another by the values of the constants as well as in the forms of the analytical expressions. Thus the information which must be added to the fundamental equations represents the condition for the solution to be unique. However, the question of how this condition should be set up and what information it should convey must remain open. The mathematical problem of the uniqueness of solutions is of extreme complexity, and its analytical study can be carried to completion only in a few quite elementary cases. This does not make it impossible, however, to present some general arguments of a physical nature. With this in view let us consider in more detail the reasons for the individual differences of phenomena.

The quantities characterizing the physical properties of the medium in a given process play an important part in the development of the process. These physical properties are constant parameters in the problem, and must be given directly in the conditions. (If it is necessary to consider the physical properties of the medium as variables also the problem becomes considerably more difficult, and in fact we have then to deal with a new problem which will be considered later.) These quantities occur in the general fundamental equations, i.e., in equations which do not contain any information as to their values. The values of all the physical constants for a problem must therefore be defined as additional information. The geometric properties of the system must also be given, since these are often not defined in the fundamental equations. The geometric properties of a system are defined quantitatively by its dimensions, so that the dimensions of the system are also constant parameters of the problem which must be given in the conditions. Thus, in addition to the fundamental equations, the conditions of the problem must be used to define a group of constant parameters which characterize the geometric and physical properties of the system in which the process occurs.

However, the required degree of determinacy is not given by just one condition.

The actual path of a process depends significantly on how the conditions act on the system at the moment that it begins and on how the external conditions continue to act during its development. In other words, the nature of a process depends to a considerable degree on two processes: the occurrences which precede it, and which give rise to the conditions prevailing in the system; and the processes which occur in a surrounding medium but have an effect on the system. Obviously it is impossible to reach the level of concreteness for which we are aiming if the features of the process being studied are not defined, including those caused by external effects, since the fundamental equations cannot contain any information of this type. The body of information required for this is expressed in the form of additional conditions, which are usually known as the *conditions for uniqueness of solution.* The concept of conditions for uniqueness of a solution, which is introduced here in a mathematical way, has a very important physical significance. In some cases, when it is impossible to carry out an exact analytical investigation of the uniqueness of a solution, more simple physical arguments are very helpful.

As regards the exact form of the uniqueness conditions, it is clear that they can only include the information which we actually have available under the particular real conditions. It is best if the uniqueness conditions can be reduced directly to a series of values of the variables (or their derivatives); this is most convenient in solving such problems. However, we do not often have such information available, and in some cases it is necessary to accept the fact that the form in which the uniqueness conditions have to be expressed leads to equations which are no less complicated than the fundamental equations of the problem. In these cases it is also necessary to give values to some of the quantities which characterize the conditions in the surrounding medium (rather than in the system); these are usually easier to determine. These values are often introduced into the conditions as known parameters. In all cases, regardless of how the conditions for uniqueness of the solution are expressed, they introduce a series of constant values for the variables being studied, and, together with the quantities which characterize the geometric and physical properties of the system, they make up a group of constant parameters of the problem. When this group of parameters is combined with the system of equations (the fundamental equations and, in some cases, additional equations also), i.e., when the prescribed values of all the parameters in the problem are introduced into the conditions, we will have defined in a single-valued manner a unique, specific phenomenon.

In the general case, the uniqueness conditions have to provide a great deal of preliminary information about the nature of the process in an extensive class of problems, which are known as *boundary-value problems* in mathematical physics; these problems are of great interest to us and will be the main subject of study. In these problems the values of the unknown quantities must be given at the initial moment at all points in space in the region occupied by the process of interest (i.e., at all points in the system), as well as at the boundaries at all times during the process. This will be discussed in more detail later. Here we will only point out that these groups of values are termed the *initial* and *boundary* values of the variables, respectively.

Thus, to set up a problem for solution, we must have a system of equations and a group of values of the constant parameters. On the basis of these, each quantity can be determined as a single-valued function of the independent variables and the parameters.

As a result, the solution of the problem includes not only the independent variables, but also the parameters, i.e., quantities which have completely defined values for the given phenomenon, but which change from one phenomenon to another. This is of major importance here. There is no direct relationship between the unknown quantities and the independent variables: the unknown quantities have to be expressed in terms of other quantities—the parameters which define the effect of the properties of the system on the development of the process. It is impossible to isolate in pure form the relationships which interest us by studying the way in which the unknown quantities change as functions of some of the independent variables, since these include the effects of many different factors which are expressed by means of the corresponding parameters.

Thus the complexity of the equations giving the laws of the process (i.e., the equations representing the solution to the problem) is caused by the fact that they include a large number of parameters which have to be introduced into the equations. The large number of parameters (which is an unpleasant feature, but which is typical for this type of problem) greatly complicates investigations of the present sort. In particular, it leads to great difficulty in presenting the results of a numerical solution or of experiments.

2. THE CONSTANT PARAMETERS OF A PROBLEM

The parameters of the problem must appear in the final solution, since they are used to express various effects, such as the properties of the system, the external conditions, etc., which influence the development of the process. Each of the parameters is regarded as a separate independent variable which assumes a new value when the conditions change. This explains the large number of parameters. In actual fact, however, a process does not have separate effects of this kind, but specific effects of a complicated physical nature. No matter how the conditions are changed, the course of the process in the final analysis is determined by the intensity of these effects. Unfortunately, as we pointed out earlier, the investigation cannot be carried out in terms of quantities which correspond directly to this type of physical effect, which is characteristic of the process. (This would have obvious advantages, since it would lead to fewer parameters and at the same time would make possible a clearer representation of the process mechanism.) A quantitative investigation must be set up in terms of the primary variables in such a way that each of these quantities expresses the effect of one of the factors which as a group determine the intensity of the effect being studied. Thus the effect as a whole is not clearly indicated by the equations we have to use: such effects are inevitably fragmented into a mass of separate factors with the usual results. On the other hand, it is quite clear that the physical substance of a problem cannot depend on the type of quantities in terms of which it is considered. When we make the transformation into the primary variables, these serve only as a means for expressing the effects which are characteristic of the process; these effects retain their importance for the development of the process regardless of whether they are expressed in an explicit equation or whether they cannot be expressed directly.

Thus the large number of separate quantities which appear as independent variables is due not so much to the physical nature of the process being investigated as to the nature of the methods of investigation which have to be used. The problem now arises of how to use the primary quantities most efficiently. The discussion so far has shown that it is not helpful nor efficient to carry out an investigation by expressing the fundamental effects in terms of the primary quantities and to simply reduce the problem to finding the relationships between a multitude of separate variables. The difficulties arising in such an approach are not entirely due to the nature of the problem being investigated, the physical nature of which is only obscured in this way by being lost amidst the large number of different quantities. By considering the individual quantities separately, neglecting their relationships with the other quantities, it is impossible to arrive at a correct idea of the part that each plays in the development of the process. In different processes (or even in the same process) the same quantity may assume different relationships towards the other quantities, and it is upon these relationships (whose form is specified by the physical nature of the process) that the real nature of the effect depends. For example,* it seems quite natural to suppose that the rate of development of a process will depend on the dimensions of the system. However, it is characteristic that the duration of an effect caused by the motion of a fluid varies as the first power of the system dimensions, while the duration of a heating (cooling) process in a body varies as the square of its dimension. In this case we have compared the relationship between the duration of a process and the system dimensions for two different processes. Let us consider another example. The nature of the motion of a fluid depends on the relationship between three forces which act on it: the force of gravity, the force of internal friction, and the force of inertia. The ratio of the force of inertia to the force of internal friction varies as the fluid velocity and the first power of the system dimension, while the ratio of the force of inertia to the force of gravity varies directly as the square

---

*All the examples mentioned here will be considered in more detail later.

of the velocity and is inversely proportional to the first power of the dimension.

These examples illustrate the crux of the matter. Each of the quantities represents one of the factors under whose influence the process develops. However, each of these influences does not act separately, but in conjunction with others in well-defined combinations in the form of complex effects; it is only these effects which are important in the development of the process. Consequently, the separate influences do not exist as such, but only their overall effect, and so the separate quantities are not important, but only their combinations which correspond to these effects. The final conclusion of this argument is that *the primary variables must be introduced not as isolated, individual parameters, but in the form of groups,* the structures of which are specified by the interactions of the various influences. These groups can also be regarded as variables or, more exactly, as parameters, which change with the physical conditions, and which make it possible to consider the problem from the point of view of the particular process being studied.

We have arrived at the very important conclusion that the important parameters for any given process are not the separate primary quantities, but certain groups of these which are completely defined for the process being considered. It is only by grouping the primary variables according to rules obtained by means of a quantitative investigation that an adequate physical model of the process is obtained. Thus, each problem must be considered individually to find its variables, which will be made up of the primary quantities.

This is the direct answer to the problem of the form in which the primary quantities must be used. However, a new problem arises directly—the method of constructing the groups. It follows

from general considerations that there must be a definite relationship between the structure of the groups and the structure of the
equations which are obtained on setting up the problem. Actually,
all the primary quantities in terms of which the problem is expressed occur in these equations in the expressions which define
the physical effects characteristic of the particular process. The
influence of each quantity on a given effect is determined by the
structure of the expression corresponding to this effect. However,
the development of the process as a whole depends on the relative
intensities of the individual effects. Thus, by combining these
expressions into the form of such a relationship (i.e., by considering the expressions for the specific quantities together with their
effects), we can see directly that the primary quantities must be
introduced into each combination that is specified by the nature
of the process. However, the groups are still far from being
defined by this. The expressions for the effects being investigated
must remain in force regardless of how the quantities appearing
in them are changed. These expressions consist of differential (or
more complicated) operators which give the relationship between
the various dependent and independent variables. In contrast, the
groups are made up of parameters, i.e., quantities which can be
regarded as constants for the given specific problem, and which
vary only on passing from one special case to another (i.e., on
replacing one numerical value of the problem conditions by another). Generally the operators are not applicable to such quantities
in the form in which they appear in the equations. Therefore a
ratio of operators is set up which is applicable to expressions
which are by no means groups. At the same time, these expressions must define not only the number and nature of the quantities
appearing in the groups, but also in some way (which we have not
yet defined) the structure of the groups, since in fact the operators

and groups express the very same physical concepts. The problem consists in investigating the nature of this interrelationship and finding a basis for establishing the methods for passing directly from the expressions determined by the structure of the equation to the corresponding groups, which consist of parameters. We will now proceed to the study of this problem.

Chapter II

GENERALIZED VARIABLES

3. THE RELATIONSHIPS BETWEEN CONSTANT PARAMETERS AND OPERATORS

Let us consider an equation of the form

$$D_1 + D_2 + \dots + D_r = 0, \tag{2.1}$$

which gives the relationship between the variables x_1, x_2, ..., x_n. Here D_1, D_2, ..., D_r are operators, each of which defines some physical effect which is of importance in the process being studied.

Equations of the type (2.1) are characteristic of the problems which are of interest to us. Cases which do not fit into this general type will be considered later (see Chapter IVA).

Following the scheme mentioned earlier, we can set up the ratios

$$\frac{D_i}{D_k} = d_{ik}.$$

If there are r operators in all, $(r-1)$ such independent ratios can be set up. (The case $i = k$, corresponding to $d_{kk} = 1$, is not considered.) The problem facing us consists in finding methods of passing directly from the operators d to groups made up of parameters (i.e., of physical constants and values of the variables specified by the conditions of the problem).

The operators and the groups are closely interrelated in the sense that they express the same facts about the physical mechanism of the process. Each operator corresponds to some group, since the operator and the group corresponding to it contain the same quantities: in the operator, these are in the form of the variables x_k, and in the group, in the form of the parameters x_{k_0}.

In addition, there must be a still closer relationship between the operators and the groups, since they must express in the same way (in agreement with the assumed model of the process) the effect of each of the factors which together constitute the physical effect being considered. However, one should not expect them to have exactly similar structures, since they play quite different roles. The operators d express the whole pattern of the process development. They include in their composition the values of variables, and vary themselves as functions of the independent variables. For any given values of the independent variables (for instance, at a given moment in time and at a given point in space) they have fixed values, *which establish the relative intensities of the corresponding physical effects.* Thus the structure of the operators must be such that in principle they can not only define correctly the nature of the effect of each quantity but can also be used for all the necessary calculations when the quantity changes according to some specific law. It is obvious that this is practicable only if differential (or more complicated) operators act on the variables. In contrast, the groups consist of quantities whose values are determined by the conditions of the problem. In any given case, the group plays the part of a constant parameter. Only one requirement affects its structure: it must express the effect of each of the quantities of which it is composed in exact agreement with the overall concept of the process mechanism. The specific laws governing the changes of the variables, which give the special features of the case being studied, can have no relation to the groups.

Thus, generally speaking, the important difference between the operators and groups is as follows. The laws of the formation of physical quantities from the primary quantities are expressed in pure form in the structure of the groups. In the operators these same laws are expressed only in a very indirect and disguised manner. This is due to the fact that the operators not only express the general laws for forming the quantities, but also define the actual calculational procedures to be used for all the special forms of the relationships between the primary variables. The structure of the operators is so complicated that it is impossible to find directly from them the laws for the formation of the quantities. As a result, it is impossible to pass directly from the ratios of the operators, which are known on setting up the problem, to the corresponding groups. It is only in the special case when the quantities which are defined by the operator are known to be constant that all the computational difficulties disappear, and both functions of the operator (determination of the law of formation of the quantities and definition of their method of calculation) actually coincide. Thus under these conditions there can be no differences between the structure of the operators and groups. This is a very interesting result, since it indicates a path to the solution of our problem. In this connection, let us analyze the following simple example.

Suppose that the quantity z is given in terms of x and y by

$$z = \frac{d^m y}{dx^m}.$$

The extent of our information about the quantity z given in the equation can be expressed as follows. The quantity z is a variable, depending on x. The nature of this dependence is determined by the form of the function $y = f(x)$.. If the function f is known, a value of z can be obtained for each value of x. By expressing the quantity z in terms of x and y by means of the derivative $\frac{d^m y}{dx^m}$, we have given the properties of this quantity so completely that it should be possible to determine its value for any way in which y may vary as a function of x. Here we have to deal with a differential operator (the simplest type), which specifies a known computational operation which has to be performed on the variables. The structure of the operator corresponds to this particular use. Now, however, let us use the equation defining z for other purposes. For our purposes it is entirely unnecessary to have the complete information required for calculating z, or to know the actual properties of the derivative $\frac{d^m y}{dx^m}$, which will depend on the form of the function $f(x)$. From the equation we are trying to find, in a general fundamental form, an answer to the problem of what influence the quantities $\bar{x}$ and y have on the effect which is defined quantitatively by the quantity z. Thus, ignoring all the information contained in the equation, we wish only to establish the law by which a quantity of the type z is constructed from quantities of the type x and y.

To solve this problem, we must consider the conditions under which z will assume constant values. Obviously it is only necessary to substitute $y = ax^m$. In this case, $z = m!a$. However,

$$a = \frac{y}{x^m},$$

where x and y are any corresponding values of the variables, in particular, the parametric values x_0 and y_0 given by the conditions. Hence the factor a can be defined as $a = \dfrac{y_0}{x_0^{\prime a}}$. There-

fore

$$z \backsim \frac{y_0}{x_0^m},$$

where $\backsim$ indicates proportionality.

Thus the law by which the quantity z is formed from quantities of the types x and y is that z is defined as a quantity proportional to the first power of y and inversely proportional to the mth power of x. This relationship, and it only, is important in solving the problem of the structure of the group corresponding to the derivative $\dfrac{d^m y}{dx^m}$; it also includes the parameters (x_0, y_0) given as conditions.

It is appropriate to note that new physical quantities are introduced in this form in the course of the investigation. Thus, the concept of velocity is reduced to the simplest case of uniform motion, and in primary terms the velocity is defined as a quantity which is proportional to the first power of the distance traveled and inversely proportional to the time which elapses $\left(v \backsim \dfrac{l}{t} \right)$. It is only when this concept is extended to the case of arbitrary motion (when l can vary with time in any given way) that it becomes necessary to replace this simple ratio of quantities by the first-order derivative $\left(\dfrac{dl}{dt} \right)$. It is clear that both expressions $\left(\dfrac{l}{t} \text{ and } \dfrac{dl}{dt} \right)$ express the same law for the formation of a velocity (a quantity of the type v) from the distance traveled and the time (from quantities of the type l and type t). However, the second of those also defines the operation by means of which the quantity v must be calculated in all cases for any specific variation of l as a function of t. Similarly, the concept of acceleration can be referred in primary terms to the case of uniformly accelerated motion, in which case it is a constant (for the given case of motion) proportional to the distance traveled and inversely proportional to the square of the time $\left(\frac{l}{t^2}\right)$. Further development of the concept leads to the definition of acceleration as the second derivative of distance with respect to time $\left(\frac{d^2l}{dt^2}\right)$. Again in this case we have two ways of expressing the same law of formation of the new quantity from those known before.

Thus the derivative $\frac{d^m y}{dx^m}$ can be replaced by the complex $\frac{y_0}{x_0^m}$. This substitution gives the essence of the transformation by means of which we can pass from the operators d to the corresponding groups. The operation of passing from the operators to the groups is termed reduction. This will be written in the form $D_i \rightarrow \Pi_i$ and, correspondingly, $d_{ik} \rightarrow \pi_{ik}$ [in our example, $\left(\frac{d^m y}{dx^m} \rightarrow \frac{y_0}{x_0^m}\right)$, where Π

(or π) denotes the group]. This operation is notable for its great simplicity, and nothing need be said on the technique of carrying it out. The formal apparatus corresponding to the arguments given here can be developed with no difficulty.

Let us return to the simplest differential operator in the form of the derivative $\frac{d^m y}{dx^m}$. It will be assumed that the function $y = f(x)$ may be given quite arbitrarily. Suppose that as well as $f(x)$ some function of x is introduced which corresponds to a constant value of the derivative over the whole range of variation of the argument. It is clear that a power expression with an exponent equal to the order of the derivative possesses this property. To distinguish it from the real law giving the variation of y in terms of x, defined by the function $f(x)$, we will term the variation given by the power

function *the fictive distribution law of y in terms of x*. In our case
(derivative of the *m*th order), we must obviously assume $y' = ax'^m$.
(The primes are used to indicate that here the quantities do not
refer to the real but to the fictive relationship.) The multiplying
factor *a* is chosen so that at a given value of the argument $x'_0 = x_0$
we will have the corresponding value of the function $y'_0 = y_0$. The
values of the variables at the beginning of the interval are assumed
to be zero (i.e., the variables are considered as though their
values at the beginning of the interval were zero). Thus

$$\frac{d^m y'}{dx'^m} = m!a.$$

However,

$$a = \frac{y'_0}{x'^m_0} = \frac{y_0}{x^m_0}$$

and, as a result,

$$\frac{d^m y'}{dx'^m} = m! \frac{y_0}{x^m_0}.$$

Thus the derivative in which we are interested is found as a
constant quantity, known directly from the conditions, in the case
of the fictive relationship. Actually it varies as a function of *x*
according to a law which depends on the form of the function *f*,
which may be very complicated. It is obvious that the constant
which has been obtained does not give either the exact value of the
derivative or the laws of its behavior. The correct view of the
nature of the relationship between the two quantities is that the
derivative obtained from the fictive relationship defines the order
of the value of the real derivative; i.e., it gives some overall idea
of the value. The fictive derivative can be used as a basis only for
making estimates of values, never for exact determinations.
The true value of the derivative is obtained from the constant

quantity derived in this way by multiplying it by some factor which is a function of x. In consequence we can assume

$$\frac{d^m y}{dx^m} = n(x)\, \frac{y_0}{x_0^m}\,,$$

where $n(x)$ includes the constant $m!$

This equation gives the exact way of writing the reduction operation which was represented earlier in symbolic form. The multiplier $n(x)$ may be a function of x of any complexity. However, it is easy to show (and this is very important for all of the subsequent discussion) that it is completely defined by the law giving the variation of y as a function of x, and is completely independent of the absolute values of the variables. Actually the last equation can be written in the form

$$n(x) = \frac{d^m \left(\dfrac{y}{y_0}\right)}{d \left(\dfrac{x}{x_0}\right)^m}\,.$$

However, this indicates that the multiplier n is identical with the derivative reduced to dimensionless form (i.e., the derivative taken of a ratio of quantities with respect to a ratio of quantities).

This representation of the properties of the multiplier n as a quantity depending on a dimensionless distribution law (regardless of the absolute values of the variables) can be derived in a completely explicit form. It is thus quite obvious that to determine n it is sufficient to take values of the argument and function which are expressed as fractions of their values at the end of the interval (or as fractions of any other pair of corresponding values).

Thus the multiplier for converting the constant derived for the power function into the true value of the derivative is completely defined in a single-valued manner by the law of the dimensionless distribution:

$$Y = f(X),$$

where $X = \dfrac{x}{x_0}$ and $Y = \dfrac{y}{v_0}$ are the relative values of the variables.

This distribution can be shown graphically (Fig. 1) as a curve on the XY plane which joins the points A (0, 0) and B (1, 1). It is very important that the quantities x_0 and y_0 are not defined in any form and can be chosen quite arbitrarily. By giving them various values, we obtain an infinity of curves which possess the general property that on transformation to dimensionless form they all reduce to a single curve, which belongs to the same family and includes the values $x_0 = 1$ and $y_0 = 1$. All these curves are obtained from the dimensionless distribution curve (and, likewise, from one another) by means of simple transformations—by multiplying the whole abscissa or the whole ordinate by constant factors (the transformation factors k_x and k_y). This means that the curves of the family being considered are similar to one another, though here the correspondence between the curves is of a wider type than an elementary geometrical similarity, since in general the factors k_x and k_y are not equal.

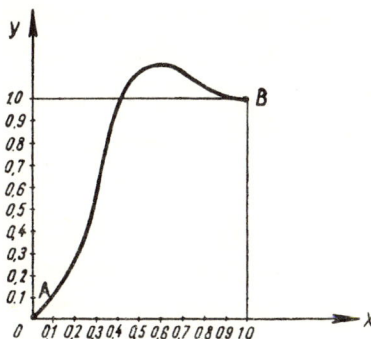

Fig. 1

Using this concept of similarity, we can see that the given function $n(X)$ corresponds to the particular family of similar curves (and, inversely, that the given family of similar curves corresponds to a unique distribution of n with respect to X).

We are now ready to turn to the transformation of an operator d as a whole. The following preliminary remarks must be made. We have regarded y as the function of a single variable, but the discussion remains valid when dependent variables which depend on several independent variables are investigated. The only change necessary in this case is that the multiplying factor n must be regarded as a function of all the independent variables. It should also be noted that our discussion is based on the assumption that it is possible to give the values of the variables at the beginning and end of the interval. This statement of the problem corresponds to the most typical case, but is not the only possible scheme for setting up the conditions for uniqueness of solution. We will show later by specific examples how to arrive at power distributions for more complicated cases which are in agreement with the real distributions (see, for example, the problem of the unsteady-state temperature distribution in a solid body).

Now let us pass on to the transformation of operators. Obviously, using the procedure just developed above, which represents essentially the operation of reduction, we find directly that $D_i = N_i \Pi_i$ and $D_k = N_k \Pi_k$. In exactly the same way, we find for their ratio

$$d_{ik} = n_{ik} \pi_{ik}. \tag{2.2}$$

Here Π denotes the group produced by substituting the fictive for the real distribution, and obviously $\pi_{ik} = \dfrac{\Pi_i}{\Pi_k}$. It is clear that the quantity Π, which corresponds to a differential operator, represents a power combination of the parameters given by the conditions. It can therefore be represented in the form

$$\Pi_j = x_1^{a_{1j}} x_2^{a_{2j}} \ldots x_n^{a_{nj}}. \tag{2.3}$$

The values of the exponents are determined by the structure of the operator D_j. Some of the exponents may be equal to zero, which means that the corresponding variable does not occur in the operator D_j. The operator D_j which is being transformed and the group Π_j obtained for it must be quantities of the same dimensions (the preceding discussion leaves no doubt on this point). However, in this case any two groups Π_i and Π_k, no matter how different in structure, must have the same dimensions (because of the similarity of the sums making up the left-hand part of the fundamental equation). As regards the relative quantities d_{ik} and π_{ik}, it is obvious that they must be dimensionless in all cases.

The multiplying factors which transform the complexes into operators are interrelated by the obvious relationship $\dfrac{N_i}{N_k} = n_{ik}$. As mentioned earlier, they are identical to the operators expressed in terms of dimensionless variables. However, they are also functions which depend only on the distribution law of the variables (i.e., on the form of the relationship between the variables, expressed in dimensionless form) and are completely independent of the absolute values of the quantities (i.e., of the parametric values which are given by the conditions for uniqueness of solution).

4. INVESTIGATION OF PROBLEMS IN TERMS OF RELATIVE VARIABLES AND DIMENSIONLESS GROUPS. GENERALIZED TREATMENT

The operators d_{ik}, which play such an important part in our discussions, can be expressed as the products of some multipliers n_{ik} with the power groups π_{ik}. Let us examine both factors in more detail and try to find the part which each plays in our system of investigation.

The groups π_{ik} represent very simple power expressions. However, this simplicity is only superficial. Actually the principles of their construction embody a deep and important idea, namely, that *the physical model of the process must be indicated in the grouping of the quantities forming the complex.* The quantities π_{ik} express the information which refers to the process as a whole. They characterize the overall properties of the process laid down by its mechanism. Specific special features of the process which occur entirely during its development are expressed by the multiplying factors n_{ik}. It is very important that these detailed properties be expressed in the form of relative distributions which do not include any absolute quantitative information. All the information which is available on the absolute values of the quantities which are important for the process must be concentrated into the groups π_{ik}, which involve the quantities and not their ratios.* It is natural that the groups π_{ik}, which correctly define the most general and important properties of the process, should be taken as the new specific quantities, replacing the various individually chosen parameters.

Thus, by means of a quantitative description of the process, the groups π are also based on a distribution of the variables in dimensionless form. Consequently, the part played by the quantitative characteristics passes to the dimensionless quantities: the parameters are introduced in the form of dimensionless combinations π, and the variables in the form of the ratios of the variable values to the values given by the conditions. This feature of the method of investigation which we are considering is very

---

*This is not at all at variance with the fact that the groups π_{ik} are dimensionless quantities. Actually the operators d_{lk} represent the ratios of the intensities of two similar physical effects and are therefore always relative quantities. However, these effects are expressed in terms of the primary quantities, which also occur as factors in the complexes π_{lk}; these factors are given in terms of absolute quantities.

fundamental and is the result of the whole series of concepts above. Actually, at the beginning of our discussion, when we introduced the operators d as the quantities defining the quantitative properties of the process (where these operators are the ratios of the intensities of the physical effects being considered), we set out on a path which must inevitably lead us to this method.

The transition to the new dimensionless quantities involves radical changes in the nature of the whole analysis. By excluding the primary quantities from consideration as independent variables, we lose that degree of determinacy which is attained by their use and which corresponds to actual phenomena.

Actually, as we have shown, a special phenomenon can be distinguished from a whole multiplicity of phenomena of the same class only by the conditions which fix the values of certain groups of quantities. However, such an operation is basically impossible within the limitations of the system of investigation being considered, since none of the quantities, taken separately, can be defined in terms of absolute values. Giving the value of the group π, and, if this is necessary from the physical content of the problem, the distribution laws in relative form at the initial moment and at the boundaries of the system, is the limit of the determinacy which can be attained with the new concepts. It must be noted that there are some individual cases which cannot be given more accurately. However, a given value of the group π may be realized by an infinite number of different combinations of the quantities forming it. Each of these combinations is specified by the value of the parameter which serves to define the phenomenon in a single-valued manner. In consequence, if the problem is set up and investigated in terms of the new ideas, an infinite number of variants of the solution is possible. Each of these variants corresponds to some actual phenomenon. Thus, what has been defined as a special individual case becomes an infinite number of different special cases.

The actual significance of this is that the new quantities are generalized and impart a generalized nature to the whole investigation. The use of the primary quantities makes it possible to set up a quantitative model of a single specific phenomenon. The new method of investigation, based on the concept of dimensionless groups and relative distributions, leads to a more complicated representation.

In addition to the concept of a unique phenomenon, we now have a new physical form—a generalized individual case. The characteristic feature of this form is that it is obtained as the highest possible degree of individualization (under the new concepts) (i.e., it is distinguished from all the other cases given by the generalized representation) and at the same time includes an infinite number of different phenomena. It must be noted that this is not mutually contradictory. It only means that the method of investigation which has been used (power groups and dimensionless distributions) is inadequate for detecting the differences which are peculiar to the phenomena included in the given generalized case. Clearly this very large number of phenomena actually exist with their own special features, which serve as the criteria of the differences and can be distinguished, as long as we are using the primary variables. However, on passing to the generalized quantities, it is no longer possible to specify these features. The phenomena become indistinguishable and merge into a single general case.

The concept of a generalized individual case is very important in the following discussion. As a result, we will now consider it in more detail. It will be regarded as obvious that the generalized individual case represents a concept which is considerably narrower than a cleass. For a given property of a class (in other words, a property which to some degree is characteristic of all the phenomena corresponding to the given system of fundamental equations) it is only possible to give groups in a general form, i.e., to define their number and structure. To isolate the generalized individual case, it is necessary to fix the numerical values of all the groups, i.e., to give the values of the parameters in generalized form. In addition, if the conditions for uniqueness of solution are such that they include values of the unknown variables at the initial moment and at the boundaries of the system (i.e., if we need information about the initial and boundary values), the corresponding distributions must also be given in relative form. These data express all of the additional information whose combination with the fundamental equations isolates the generalized individual case within the class. In this sense they are analogous to the conditions for uniqueness of solution. It is easy to see that here we are dealing with something rather larger than a simple analogy, and that the requirements being considered represent *a generalized form of the uniqueness conditions.* It is necessary to show that the combination of these requirements with the fundamental equation separates from the class a unique (single-valued) generalized case. This means that the following proposition must be proved: *the combination of the fundamental equations and the additional conditions being considered must lead to a solution in which the unknown variables, expressed in relative form, are given as* single-valued functions *of the relative independent variables and the groups* (as constant parameters).

Suppose that the process being investigated is defined by Eq. (2.1):

$$D_1 + D_2 + \dots + D_r = 0.$$

Suppose also that in order to specify a particular phenomenon we are given in addition the values of a series of quantities x,

$x_{2_0}, \ldots, x_{n_0}$ and also the unknown variable at the initial moment and at the boundaries of the system. (A boundary-value problem is considered, as a more complicated case.) In this way a mathematical model of the specific phenomenon is set up in the form of equations which are given together with a set of quantities known from the conditions. (Nothing is changed if we consider a system of interrelated equations instead of a single equation.) Now let us separate the generalized case from the class corresponding to Eq. (2.1). To do this we must convert the quantitative relationships into relative form. Let us introduce the relative variable $X_k = \frac{x_k}{x_{k_0}}$. Similarly, the information given by the field conditions is replaced by dimensionless distributions, and Eq. (2.1) is rewritten in terms of relative operators, for example, in the form

$$1 + d_{21} + d_{31} + \ldots + d_{r1} = 0.$$

The operators d are also taken to operate on the relative variables X. Thus

$$1 + n_{21}\pi_{21} + n_{31}\pi_{31} + \ldots + n_{r1}\pi_{r1} = 0.$$

The equation has assumed a more complicated form. The groups π appear in it as numerical multipliers of the operators, which are applied to the relative, rather than the absolute, variables. If the values of the groups can be chosen arbitrarily, we would have an equation with undefined coefficients; i.e., in essence it would not be a single equation at all, but an infinite number of different equations. Actually, in setting up the problem the values of all the groups must be fixed precisely, and in consequence we have only a single, completely defined equation to deal with. Thus we have obtained an equation which gives a relationship between the variables $X_1, X_2, \ldots, X_n$ together with uniqueness conditions expressed in terms of the same variables X (in the form of dimensionless distributions). Under these conditions, the solution must give the unknown variables as single-valued functions, since all the variables are relative numbers X and all the parameters are dimensionless groups π. This is what we set out to prove.

Thus we have proved the single-valued nature of the relationship corresponding to the generalized individual case, as well as the correctness of the concept of generalized conditions for the uniqueness of the solution. However, the proof contains a hidden assumption which requires discussion. Actually, in the course of the discussion the following assumption has been made tacitly as being something obvious: *if the problem is correctly set up initially in terms of the absolute variables, it will remain validly stated after transition to the relative variables.* More specifically, if the conditions for uniqueness of solution are set up correctly, they will be transformed automatically to the generalized uniqueness conditions on transition to the relative variables; i.e., they will be transformed to uniqueness conditions which correspond correctly to the new formulation of the problem. This assumption is fairly obvious from a physical point of view, since it can scarcely be supposed that the content of the uniqueness conditions should be changed by the transition to the relative variables.

5. THE GENERALIZED INDIVIDUAL CASE AS A GROUP OF SIMILAR PHENOMENA. CRITERIA OF SIMILARITY. GENERALIZED EQUATIONS

Now let us go on to solve the problem of how the concept of a specific unique phenomenon corresponds to that of an individual generalized case. It is obvious that there are two closely inter-related sides to this problem, and that both must be considered. First we have to explain the relationship between the generalized

individual case and the phenomena which are represented in it.
How do the phenomena merge into the generalized case, and how
can they be isolated from a given generalized case? In addition,
we have to establish the interrelationship between these phenomena.
What features do they have in common, and in what respects are
they different?

The discussion earlier leads us to the following conclusions.
Any phenomenon can be reduced to an individual generalized case
if its quantitative properties are expressed in relative quantities.
It is significant that on passing from absolute quantities to the
relative constant parameters which define the properties of the
phenomenon given by the conditions, they are automatically combined into dimensionless complexes and excluded from consideration as independent factors. Conversely, on passing from relative
to absolute characteristics the generalized case is transformed
into one of the phenomena which it represents. The absolute characteristics (x_k) are obtained from the relative ones (X_k) by multiplying by the corresponding parameters (x_{k_0}), thus removing them
from the groups which are thereby broken down into separate
independent quantities. Thus the two specific features of the
generalized form of analysis—the relative nature of the variables
and the complicated nature of the parameters—are closely interrelated. The actual values given in the form of the product
$(x_k = X_k x_{k_0})$ depend on how the group π is broken up into the factor
(x_{k_0}).

Each variant of the calculation gives its own numerical value
and, correspondingly, a phenomenon which is completely defined
and different from all the other phenomena included in this generalized case and produced by the other variants. All these phenomena
are closely related. The nature of this correspondence is explained
by the fact that in all cases the ratios between the intensities of the

primary physical effects are the same. All the quantitative features are identical when expressed in relative form. The boundary distributions of all of the variables are similar. This extremely characteristic form of conformity between different phenomena is defined as complete physical similarity.

Thus the whole vast number of phenomena which correspond to a generalized individual case forms a group of similar phenomena. This concept can be visualized more clearly if it is expressed in another form. As we have just explained, the quantitative characteristics of the phenomena do not coincide only because of the different values of the parameters x_{k_0} (which play the part of multiplying factors and which change from one phenomenon to another, though they have completely fixed values for each given phenomenon). The only feature serving to distinguish the distribution of the variables under the conditions of a specified phenomenon from the distributions for any of the other phenomena of the same group is its scale, which is determined by the value of the appropriate parameter. As a result, each specific phenomenon is distinguished from the group of similar phenomena by a set of scales, which are given in the form of the parameters x_{k_0}. In this sense, the various phenomena of the same group can be regarded as *one and the same phenomenon given in terms of different scales.*

Thus the generalized individual case is a relative form of representing a group of similar phenomena. A definite set of numerical values of the groups π corresponds to each generalized case, and this is its unique quantitative feature. Therefore the groups have the same values for all the phenomena included in the same group. These phenomena do not have to satisfy any other requirement. This means that the numerical values of the groups corresponding to the relative operators completely define the quantitative properties inherent in the group of similar phenomena and characterizing the group as a whole. As a result the equality of these groups is a necessary and at the same time unique (that is, sufficient) characteristic for the similarity of the phenomena. It is obvious that this feature can be used as a criterion for establishing whether phenomena are similar. In this connection the groups π can be legitimately termed similarity criteria. It should be noted that, according to the principles on which the groups π are constructed, each of the criteria can be regarded as some mean measure of the ratios of the intensities of two physical effects which are important in the process being studied. Thus the numerical values of the criteria can be used as a basis for overall quantitative evaluations. The accuracy of such estimates increases the more completely the characteristic inherent tendencies due to the mechanism appear in the development of the process. They also become more exact as the effects being compared become more constant in intensity.

The general concept leading to the idea of the individual case as a model for a multitude of similar phenomena points the way to new possibilities in quantitative investigations; these arise as a result of the transition from special to generalized relationships. A specific phenomenon, whose study leads to quantitative relationships, can be regarded as representative of a very large number of phenomena which constitute a generalized case. Accordingly, relative variables are introduced, and all the constant parameters are replaced by groups. The quantitative results of the investigation are represented in the form of dimensionless relationships, in which the unknown variables (expressed in the form of ratios) are defined as functions of the independent variables (also in relative form) and the similarity criteria. All the earlier steps have been leading up to this one. Actually, any quantitative relationship obtained for some specific phenomenon can be extended to all the other phenomena similar to it if they are reduced to the relative form: *an identity in the relative form is transformed into similarity on transition to absolute quantities.* This defines the role of similarity criteria in our system of investigation.

We have now arrived at a very rational form for expressing quantitative relationships which has two important advantages in use. The first is that the number of arguments is decreased, since the parameters are grouped in the form of similarity criteria. The second is that the applicability of the results which are obtained is greatly increased. However, certain problems must be cleared up in advance in order to be able to use this form efficiently in practice.

6. COMBINATION OF CRITERIA AND RELATIVE VARIABLES. CRITERIA OF THE PARAMETRIC TYPE. THE FORM OF DIMENSIONLESS VARIABLES

The forms of the groups which occur in the final equations are not determined strictly. To a certain degree they can be constructed arbitrarily (since the pairwise combination of the operators to be compared can be carried out in any way). In addition, the following must be remembered. As we have explained, the numerical values of the similarity criteria are unique quantitative features of the generalized case, and, correspondingly, equality of the criteria is the only prerequisite for the similarity of the phenomena which have this quantitative content. This determines the role of the similarity criteria and their place in the system of generalized ratios. Suppose that two criteria π_1 and π_2 are obtained on evaluating an equation. The requirement that they must have specified numerical values (i.e., that they must have values which are the same for all the mutually similar phenomena) can be replaced by any equivalent requirement, for instance, that the expressions $f_1(\pi_1, \pi_2)$ and $f_2(\pi_1, \pi_2)$ should have specified values (in other words, that these should be equal for all the phenomena), where the functions f_1 and f_2 are chosen quite arbitrarily. This leads to the following conclusion: *any combination of similarity criteria is also a similarity criterion.*

Let us consider further the validity of combining criteria and relative variables. It is obvious that multiplying the relative variables (whether the independent or the unknown variables) by a fixed constant factor cannot have an effect on the single-valued nature of the relationships which are obtained. However, for each generalized case any combination of the criteria is in fact a fixed constant factor. As a result, *the product of a relative variable and any combination of similarity criteria is also a similarity criterion.*

Finally, it must be noted that the possibility of combining the criteria and relative variables is very important as the basis for a rational construction of the criteria themselves. Actually, in carrying out the reduction operation we find all the time that different ranges of variation must be introduced for the same quantity. Thus, if some variable u appears in an operator once in the form of the derivative $\dfrac{\partial^m u}{\partial x_i^m}$, and again in the form $\dfrac{\partial^n u}{\partial x_j^n}$, reduction gives $\dfrac{u_i}{x_{i_0}^m}$ and $\dfrac{u_j}{x_{j_0}^n}$ respectively. The numerators of these expressions are not equal (note the different superscripts), since they represent changes of the function corresponding to definite ranges of change of two different variables (for instance, a change in the quantity being investigated over a given length and the change in the course of some given interval of time). Thus the group must contain different values of u. The disadvantages of this are obvious. However, this difficulty can be easily eliminated.

Combining the group with a factor with multipliers $\frac{u_i}{u_0}$ and $\frac{u_j}{u_0}$ reduces it to a form in which the variable u can be represented parametrically only by the value of u_0.

The properties of the similarity criteria which are being considered here are very useful because in practice the use of generalized relationships frequently makes it necessary to combine criteria to eliminate difficulties which arise from various causes. The two cases which we will consider now are very typical in this respect. They refer to certain peculiarities in the way which the conditions of the problems are set up. Their discussion is very instructive and will lead to important new ideas.

Frequently the conditions of a problem are formulated in such a way that it is necessary to consider not one but two or more values for some quantities (for example, the diameter and height of a cylinder; the frequency of a natural oscillation and the frequency of an external disturbance; the absolute velocity of a medium, its relative velocity, and the velocity of propagation of a disturbance in the medium, etc.). In other words, the conditions must contain two (or more) parameters of the same physical type. It is obvious that in this case we obtain criteria which are completely identical in structure, but differ in having different values for at least one of their constituent quantities. By combining such criteria in pairs they are replaced by ratios of parameters of the same type, and we retain in the initial form only one criterion for each group of similar groups. The criteria obtained in this way are simpler than the initial ones. There are indisputable advantages in doing this. In addition, they can be formulated directly from the conditions of the problem, without considering the equations. Thus, parameters can be introduced into the final equations not only in the form of groups but also as ratios of quantities of the same type. This type of ratio will be termed *criteria of the parametric type*, or *parametric criteria.* Quite frequently we encounter parametric criteria of a geometric nature, which obviously represent the dimensions of the system expressed relatively. It is easy to see that when these criteria are similar for different phenomena it means that the systems in which they occur are geometrically similar. Similarly, parametric criteria of a physical nature express the condition that the corresponding fields are similar. Obviously the parametric criteria characterize certain properties of the phenomena which are defined by the conditions of the problem and can be set up directly without referring to the equations. As a result, we can say that the parametric criterion is the specific form for describing some conditions of the problem.

Thus the appearance of parametric criteria is explained by the fact that more than one value of the corresponding quantity is defined in setting up the problem. Often the opposite situation arises, when the conditions contain no value for a given quantity. A characteristic (but, of course, not unique) example of this distinctive indeterminacy occurs in the case of unsteady-state non-periodic processes, for which it is impossible to define a known duration in advance.

If there is no parameter x_{k_0} in the conditions corresponding to some variable x_k, there are two obvious consequences: first, it is impossible to reduce the variable to a relative form $\left(\frac{x_k}{x_{k_0}} \right)$; second, it is impossible to formulate criteria containing the parameter x_{k_0}. This difficulty can be easily overcome by combining the variable $\frac{x_k}{x_{k_0}}$ and the corresponding criteria in such a way that the parameter x_{k_0} is excluded. It is not difficult to see that this produces expressions which agree in form with the criteria being transformed, but with the unknown parameter x_{k_0} replaced by the variable x_k. This type of expression must also be regarded as a criterion. It represents a special dimensionless form of the variables that is obtained by referring an absolute value of a quantity not to a parameter of the same nature (since it is not given), but to a group of other parameters (known from the conditions) which form a combination equivalent to the deficient parameter.

Let us discuss this problem in rather more detail. The fact that the parameter x_{k_0} is combined with a group of other parameters into a single group π has a deep physical significance. It means that taken all together they have a single effect as a whole on the development of the process, and so each of them can be replaced by a combination of the others. Thus any parameter is equivalent to a group of quantities which together with it (in the form of a ratio) form the group π and can be represented by this group under all conditions. In this sense, this is equivalent to replacing the relative variable $\frac{x_k}{x_{k_0}}$ by a different dimensionless variable. These two expressions are completely equivalent. In contrast, the criterion π and the dimensionless variable are essentially quite different in all external respects. The criterion π expresses a definite relationship between the conditions of the process, which are expressed by the corresponding parameters. This relationship is *the necessary prerequisite for similarity*, and this is the basis for calling the group π a similarity criterion.

This property is not in the least characteristic of the dimensionless variables, and as a result we have not extended the term "similarity criterion" to them; this is reserved only for *groups formed from constant parameters* (though it must be noted that in the recent literature this distinction in terminology is not observed, and the term "similarity criterion" is applied without constraint to expressions involving variables).

To sum up our discussion of this interesting problem, it can be said that in the absence of a corresponding parameter in the problem conditions it is necessary to replace the simple relative variable by a dimensionless variable, and one of the criteria drops out. It should be noted that when the missing parameter appears in several of the criteria rather than in a single criterion, it is easy by grouping them appropriately to produce products of the criteria which do not contain this parameter. It is obvious that the number of products obtained in this way is always one smaller than the initial number of criteria. However, sometimes the situation is complicated when two (or more) parameters are not defined by the conditions when these are required in the formation of the relative variables, for example, the time (nonperiodic process) and a characteristic dimension (unbounded system). In these cases, attempts to eliminate the missing parameters may lead to groups containing more than one independent variable. (See, for instance, the problem of the temperature field in a body of unrestricted dimensions.) Finally, we also meet problems in which the absence of scale values for the unknown variables and argument can be dealt with by combining them in a single group (for example, the problem of the time taken for a temperature front to penetrate to a given depth).

The general conclusion must be that the structures of the expressions appearing in the final solution are determined by the way in which the problem is stated.

7. THE POSSIBILITIES AND OBJECTIVES OF GENERALIZED ANALYSIS

We have arrived at a system of quantitative investigation which can be logically described as the method of generalized analysis. This method is based on the systematic use of dimensionless quantities—similarity criteria and relative variables.

The similarity criteria represent a twofold type of combination of the constant parameters of the problem. Criteria of the parametric type are simple ratios of parameters of the same sorts; they can be set up directly from the conditions of the problem. Criteria of the group type include different sorts of parameters. They are obtained from the equations in the form of power expressions which reproduce the structure of the dimensionless operators.

The relative variables represent the results of dividing the problem variables by constant parameters. Two types of relative variables must be distinguished (just as there are two types of criteria). The most natural and simple form of relative variable is the fraction formed by dividing its variable value by a parameter of the same sort—this is analogous to a criterion of the parametric type. If the parameter corresponding to the particular variable is not defined by the problem conditions, the relative variable is set up in the form of a group, which in form is identical with a criterion in which an unknown parameter (one not given by the problem conditions) is replaced by the value of the variable.

The solution of the problem is given in the form of an equation between dimensionless quantities in which the unknown relative variables are defined as single-valued functions of the independent relative variables and the similarity criteria, which play the part of constant parameters.

It must be noted that some of the groups π (particularly the most important ones in theoretical investigations and calculations) are normally denoted by the first two letters (or less frequently by the first letter only) of the surnames of the scientists who played an important part in the development of the particular field, and are named accordingly; for example, Re is the Reynolds criterion. Sometimes symbols of the type N_{Re} are used instead, but this form is not widely accepted.

As mentioned earlier, all the groups are usually called criteria. In contrast, we propose to use the name criterion only for those groups which are composed entirely of parameters given by the conditions. We will call relative variables of the group type *numbers*, for example, the Fourier number (the dimensionless form of the time variable in problems involving the theory of heat conduction).

Thus, this leads to the following general type of equation for defining the unknown variable Y as a function of the independent variables X and the similarity criteria:

$$Y = f(X_1, \ X_2, \ldots; \ \pi_1, \ \pi_2, \ldots; \ P_1, \ P_2 \ldots),$$

where π is a criterion of the group type;

P is a criterion of the parametric type.

All the variables $X_1, \ X_2, \ldots; \ Y$ are expressed in relative form.

It must be noted that the ultimate aim of the generalized analysis is to determine the structure of the generalized variables which are characteristic of the process being studied. The solution of this problem exhausts its possibilities. The form of the function in the final expression for the unknown variable cannot be determined by means of a generalized analysis—this is fundamentally impossible. The final answer can only be obtained as a result of solving the problem (analytically or numerically), carrying out experiments, or providing additional physical arguments. In all cases the means of carrying the solution to completion lie outside the limits of generalized analysis.

Generalized formulas of the power type $(\pi = A\pi_1^{n_1}\pi_2^{n_2}...P_1^{m_1}P_2^{m_2})$ are widely used in engineering practice. This method of representation is completely arbitrary and has no theoretical foundation. The use of power formulas must be regarded purely as a computational device.

Chapter III

SOME BOUNDARY-VALUE PROBLEMS

8. PROCESSES IN CONTINUOUS-MEDIA AND BOUNDARY-VALUE PROBLEMS

Now let us pass on to a study of processes which occur in continuous media. There are many types of these processes which are inherently different physically. Nevertheless, certain characteristic features observed in their quantitative study belong to all to some degree. This makes it possible to set up a general scheme for their quantitative study; this is determined by the way in which problems occurring in the study of processes in continuous media are set up.

The medium in which the process occurs (this may be a solid, a liquid, or an elastic fluid—a gas) is regarded as a continuous material continuum (this does not prohibit its actual physical discreteness, as long as the natural dimensions of the system are not commensurate with the mean free path of the molecules). As a result, each point in the space occupied by the process can be associated at any moment of time with definite values of all the quantities which are important for the particular process. Hence any of these quantities is characterized by a group of instantaneous values which are continuously distributed in space (such a group of values is termed the field of the given quantity).

Thus the picture corresponding to some given moment during the development of the process is represented as a system of different physical fields. Under the conditions of a steady-state process this picture remains unchanged. In the case of an unsteady-state process it changes in such a way as to express the change of the physical circumstances with time. It is obvious that the process can be regarded as completely defined if it is possible to give the fields of all the variables at all moments of time. Complete information is necessary to make it possible to derive the distributions of the variables in time and space; this is the ultimate aim of a quantitative investigation.

Thus any problem in the theory of continuous media can be reduced to the determination of certain physical quantities (the ones which are important for the process being investigated) as a function of the coordinates and time. The conditions for setting up the problem correctly must contain the complete body of initial information required for carrying the solution to an exact, single-valued result. These data are expressed in the form of the fundamental equations of the problem and the conditions for uniqueness of solution.

As we have shown, the fundamental equations are the mathematical model of the internal mechanism of the process being studied. This means that they express our ideas concerning the interactions of the elements of the medium. At the same time, they designate the course of the process, since they establish the relationship between the physical conditions at a given moment and the timewise changes of these conditions. (Consequently, the interactions between the elements of the medium are the reason for its change of state.) Hence, to the extent that the development of the process is governed by the interactions between the elements of the medium, it is described by the fundamental equations.

However, the elements of the system which exist at its boundaries interact with the bodies surrounding the system as well as among themselves. Some external effects must therefore penetrate into the system, and naturally must have some influence on the nature of the process. However, in this case the conditions for the interaction of the system with the external medium must be expressed independently of the equations. The collection of information additionally given to define sufficiently closely the conditions at the surface where the system and surrounding medium interact is termed the *boundary conditions*.

The fact that the information contained in the fundamental equations is insufficient is clear in other ways also. The equations permit us to find a succeeding state of the system from one given previously. However, they do not include any information which enables us to establish some initial state which can be used as a starting point. The state of the system at the initial moment of the process cannot be regarded as being within the scope of the fundamental equations, since this state is the result of various other processes which have occurred earlier in the same region of space. By analogy with the ideas above, therefore, we introduce the concept of the initial condition as a collection of information which defines the initial state of the system and which is given in addition to the fundamental equation. This is not just a superficial analogy, but a very fundamental one. Processes which are not being studied because they occur outside the space or time limits of the region of investigation have an effect on the development of the phenomenon being studied and therefore must be suitably expressed in terms of additional conditions.

These boundary and initial conditons, which together give conditions for uniqueness of solution for the problem being studied, are combined in the concept of *boundary conditions*, since they refer to the "boundaries" of the space-time region within which the process occurs, and problems of the present type are termed *boundary-value problems*. Thus, boundary-value problems can be stated in the following specific form: determination of the conditions within the whole space-time region of interest using the fundamental equations with specified conditions at the boundaries of this region. It must be noted that the quantities which define the physical and geometric properties of the system (the physical properties of the medium and the dimensions of the region) are of great importance in boundary-value problems. These quantities are important for the process because they define the effects of the natural properties of the system on the development of the process. This will be the case as long as we regard the quantities as constant parameters. In many cases this simplification corresponds to the practical requirements as regards the accuracy of the solution. However, generally speaking, the physical properties of a material vary with its state. Therefore, for a more exact statement of the problem the physical properties must be regarded as functions of the appropriate variables (in the first place, as a function of temperature). Let us consider this problem in more detail.

The quantitative investigation of physical problems (particularly applied problems) frequently leads to boundary-value problems, and naturally these are of particular interest to us.

Let us consider some typical boundary-value problems. It can be said in advance that in all cases the generalized solution must be sought in the form:

$$\frac{u}{u_0} = f\left(\frac{t}{t_0}, \; \frac{x_1}{l}, \; \frac{x_2}{l}, \; \frac{x_3}{l}; \; \pi_1, \; \pi_2, \ldots; \; P_1, \; P_2, \ldots\right),$$

where u is the unknown variable;

$\quad\quad\quad$ t is the time;

x_1, x_2, x_3 are the coordinates;

$\quad\quad\quad$ l is a characteristic dimension given by the conditions;

$\quad\quad\quad$ 0 is the subscript on the values of the quantities given by the conditions which play the part of scale-factors and which are used in forming the relative quantities (we have introduced the quantity l for lengths instead of x_0).

These values occur in the groups π. If several different values are given in the problem conditions for the same quantity, one of these is chosen as the reference scale (the others appear in the form of the parametric criteria P).

A. The Temperature Field in a Solid

9. THE INITIAL PHYSICAL CONCEPTS AND MATHEMATICAL STATEMENT OF THE PROBLEM OF THE TEMPERATURE FIELD IN A SOLID BODY. GENERAL FORM FOR REPRESENTING THE TEMPERATURE FIELD

First let us consider the process by which heat redistributes itself in a solid body. The investigation of this process leads to a quite typical boundary-value problem; the process itself is fairly simple, since the transfer of heat is not accompanied by the movement of any material, so that the process occurs in a stationary medium. The quantitative picture of the process is given completely by the field of only one physical quantity—the temperature (as we shall see, the heat flux is uniquely defined for a given temperature field). Therefore, there is only one fundamental equation for the case being considered; this defines the temperature as a function of the coordinates and time. We do not have to deal with a system of fundamental equations, as in many other cases. Because of the relative simplicity of the process, the problem can be stated exactly. This imparts rigor and completeness to the whole analysis. As we can see, this problem is very suitable in all respects to be considered first.

Let us set up the physical representation. Heat is redistributed in a solid body if (and only if) the elements of the solid have different temperatures. In consequence, the necessary and sufficient condition for the process to occur is that the temperature field must be nonuniform.

The heat flux passing through a body with given physical properties corresponds exactly to the temperature distribution in it. The precise form of this correspondence is given by the well-known Fourier equation:

$$\vec{q} = -\lambda\,\mathrm{grad}\,T, \tag{3.1}$$

where $\vec{q}$ is the heat flux vector, whose absolute value is the heat flux density, i.e., the quantity of heat transferred per unit time per unit area of the surface through which the flux passes;

T is the temperature;

grad as usual is the symbol of the vector operator known as the gradient;

λ is the thermal conductivity, i.e., that physical property of the material which plays the most important part in the process being studied.

Equation (3.1) expresses a definite physical hypothesis; in common with all hypotheses of this type, it can be checked experimentally. The validity of this equation has been confirmed by more than a century of practical application. It has been found that the thermal conductivity varies with temperature, as indeed do the values of all the other physical constants. Taking the variability of the thermal conductivity into account greatly complicates the problem of finding a solution. As a result, the quantity λ is taken to behave as a constant whenever this does not lead to very large errors. In what follows the dependence of λ on T will not be taken into account.

Thus the process of heat flow in a solid is completely defined by the properties of the temperature field. It stands to reason that the temperature distribution is of independent interest, if for no other reason than that all the properties of the material depend on the temperature. It is therefore valid to take as the foundation for the study of the present problem the equation defining the temperature as a function of the coordinates and time (and, of course, the physical constants applicable to the process).

Let us consider the derivation of this equation, assuming that the medium is homogeneous and isotropic. Within the solid we will isolate an element which is so small that the temperature can be regarded as constant throughout its whole volume at the given instant of time. The state of this element can change only as a result of heat transfer with the medium surrounding it, which represents the whole collection of adjacent solid elements.

Heat transfer occurs if the temperature of the element differs from that of the surrounding medium. In this case the internal energy of the element must also change, and this results in a change of its temperature.

Thus the general physical principle on which a quantitative investigation of the process can be based is the law of conservation of energy. Under the conditions being considered, when there are no effects due to the conversion of energy, and the process consists only of a redistribution of heat among the elements of a solid, this law can be reduced to the statement that the change in the internal energy of an element can be equated with the quantity of heat which it exchanges with the surrounding medium.

If dU denotes the change in the internal energy of an element during the time dt, and if dQ denotes the quantity of heat which it exchanges in the same time with the surrounding solid mass, the equation can be written in the form

$$dU = dQ$$

or

$$\frac{dU}{dQ} = 1.$$

We have arrived at an equation which is simple with regard to both its mathematical form and the sense of what it states. The whole quantitative content of the problem is reduced in this equation to the condition that the ratio of the internal energy change of the element to the quantity of heat gained (or lost) by it in the course of the whole process must be equal to unity. However, the equation is useless in this form, and does not contain any of the quantities which would allow it to be applied directly.

We are actually interested in the relationship between the primary variables, as discussed earlier. For the present problem these variables are the temperature, the coordinates of a point in the body, the time, and the physical properties characteristic of the body which are relevant to the process being studied. There is no particular difficulty involved in converting to these quantities. We can write immediately:

$$dU = c\rho dT dV.$$

After some simple calculations involving the use of Eq. (3.1), which relates the heat flux to the temperature distribution, we arrive at an expression for the elementary quantity of heat:

$$dQ = \lambda \nabla^2 T \, dt \, dV,$$

where ∇^2 is the Laplacian operator (i.e., the symbol denoting the operation of summing the second derivatives of a scalar quantity with respect to the coordinates);

dV is the volume of the element;

c is the heat capacity of the body;

ρ is the density of the body (hence the product $c\rho$ represents the volumetric heat capacity).

The fundamental equation now assumes the form

$$\frac{\partial T}{\partial t} = \frac{\lambda}{c\rho} \nabla^2 T$$

or

$$\frac{\partial T}{\partial t} = a \nabla^2 T, \tag{3.2}$$

where $a = \dfrac{\lambda}{c\rho}$ is the physical property known as the thermal diffusivity.

We have obtained the fundamental equation for the transfer of heat in a solid body. Even with the simplifications which have been made (physical constants independent of temperature, homogeneous and isotropic medium) it can be set up only in terms of a partial differential equation. It gives no direct relationship between the variables. The equation establishes only a definite relationship between the change of temperature with time at each point in the body and the distribution of temperature in the space around this point. Specifically, this states that the rate of change of temperature with time increases as the curvature of the temperature distribution (at the given instant, about the given point) increases along each of the coordinate axes. The time derivative of the temperature is connected with the derivatives with respect to the coordinates by means of a proportionality factor. This means that the rate of reorganization of the given temperature field is determined by this quantity. Consequently, we can say that the thermal diffusivity characterizes the ability of the material to respond to the passage of heat by changing its temperature.

The example being considered is an excellent illustration of our earlier discussion of the reasons leading to the complex nature of the fundamental equations. On the one hand, we have the energy conservation equation, written in terms of heat quantities, which is extremely simple both in sense and form; it merely expresses the fact that two components must be equal. On the other hand, we have the temperature-field equation. As we have already mentioned, even with the simplifying assumptions this can only be expressed in terms of a second-order partial differential equation. Such are the consequences of the transition from the concept of an energy flux to the simple primary quantities (temperature, physical constants).

The characteristic incompleteness of the information given by the fundamental equation about the process can also be seen clearly. Equation (3.2) expresses the process of heat redistribution among the elements of the solid body. Hence it is able to characterize the temperature field only insofar as this field is determined by the natural redistribution of heat within the body. However, the body being studied is not infinitely large, and it is impossible to omit from our consideration the effects caused by the passage of heat through it from outside. These effects, which arise from the interaction of the body with the surrounding medium, are caused by various other processes, and naturally are not defined in any way at all by the fundamental equations. Thus the phenomenon being investigated (the formation of a temperature field) develops as the resultant effect of various different types of processes which are occurring simultaneously within the body and at its boundaries. It should be noted that the argument is not affected when the body is thermally insulated, since then it is necessary to satisfy the condition (not contained in the equation) that there is no gain or loss of heat at the surface.

The deficiencies of the equation in other respects can also be seen clearly. If the temperature distribution at any instant is given, it is easy to find from Eq. (3.2) the value of the time derivative of temperature at any point, and thus to determine completely the temperature field for any subsequent (or previous) instant. However, without assuming an initial distribution, we cannot obtain concrete results. Equation (3.2) enables one to follow in detail the reorganization of a given temperature field, but it contains no information as to the initial field.

In considering a specific case the general remarks on the indeterminacy of the information given by the fundamental equation have achieved a clear and simple sense. Equation (3.2) must be supplemented with boundary conditions. The initial condition is given in the form of an equation defining the initial temperature field (i.e., at the moment $t = 0$ the temperature is given as a function of the coordinates). The boundary conditions can also be given in similar form (the temperature is given as a function of time for each point on the surface of the body).

This method of defining the boundary conditions is logically very reasonable and technically very satisfactory. However, it cannot be used in most cases, since the surface temperature of the body is unknown. For the same reason, it is impossible to realize the second form (possible in principle) of stating the boundary conditions, which consists in defining the heat flux density for all points on the surface (this is obviously equivalent to giving values of the derivative of the temperature with respect to the normal to the surface at all its points). Usually we are only given the temperature of the surrounding medium. The form of the boundary conditions which defines the circumstances at the surface of the body in terms of the medium surrounding it is therefore much more characteristic. This relationship between the conditions at the surface of the body and in the surrounding medium is given by the heat transfer equation:

$$\alpha \Delta T = \lambda \, | \, \mathrm{grad}_n T \, |_0. \tag{3.3}$$

Here $\Delta T \equiv | \, T_m - T_c \, |$ is the temperature driving force, i.e., the absolute difference between the temperature at the surface of the body and that in the surrounding medium;

$| \, \mathrm{grad}_n T \, |_0$ is the temperature gradient at the surface of the body in the direction normal to its surface;

α is the heat transfer coefficient, i.e., a quantity, defined as the density of the heat flux which the body exchanges with the surrounding medium, referred to unit temperature driving force.

Equation (3.3) states that the quantity of heat which the body exchanges with the surrounding medium is equal to the quantity of heat passing through its surface. This condition must be satisfied at each point on the surface. The quantities appearing in the equation (α, ΔT, $| \, \mathrm{grad}_n T \, |_0$) can vary from one point to another. In some cases they have to be taken as functions of the coordinates, which complicates the problem still more.

However, usually it is possible to introduce mean values over the whole surface for all these quantities. Hence in the solution of a steady-state problem they behave like constants, while they are functions of the time only for unsteady-state problems. The heat transfer coefficient is a quantity given directly by the conditions in all cases. Very often it is possible to neglect changes of the heat transfer coefficient with time. It therefore has the significance of a constant parameter given by the conditions.

The representation of the specific heat flux in Eq. (3.3) as a product of two quantities—the heat transfer coefficient and the temperature driving force—is widely used, particularly in the technical literature. It expresses a definite concept of the way the physical conditions of the heat transfer process affect its intensity. The difference between the temperature at the body surface and in the medium surrounding it is the actual reason for the process' occurring. This difference—the temperature driving force—therefore has a very important effect on the value of the specific heat flux. However, for a given temperature driving force, the specific flux may have various very different values depending on a large number of factors which determine the physical circumstances of the process (the properties of the medium, its state of motion, the geometric properties of the body). Obviously it is advisable to isolate the direct effect of the temperature driving force in order to obtain a quantitative measure of the intensity of the process caused only by the physical features of the interaction between the body and the medium. The heat transfer coefficient provides such a measure. In addition, the conditions of the interaction between the body and medium vary with the temperature driving force. This effect can be neglected in many important cases, and the heat transfer coefficient can be regarded as a quantity which is independent of the temperature driving force. Even if the effect of the temperature driving force is quite appreciable (a typical case, which we will consider later, is that of heat transfer under conditions of free motion), α depends on ΔT to a much smaller degree than q, and this dependence has an averaged, indirect character.

Physically speaking, the heat transfer coefficient is a complicated quantity. It expresses an overall effect which includes the effects of a number of phenomena of different types. It is quite a difficult problem to determine the actual value of the heat transfer coefficient from the given conditions of a process.

In the investigation of temperature fields in solid bodies the heat transfer coefficient must be regarded as a given quantity, as we mentioned earlier.

Hence there are three different forms for giving the boundary conditions; these are usually designated as boundary conditions of the first, second, and third types, of which the conditions of the third type are the most typical. We will consider this case.

The problem, therefore, is stated in the form of the fundamental equation (3.2):

$$\frac{\partial T}{\partial t} = a\nabla^2 T,$$

and the boundary conditions are given in the form of an equation for the initial temperature field (for example, $T = \mathrm{const} = T_0$ at $t = 0$) and Eq. (3.3):

$$\alpha \Delta T = \lambda \,|\, \mathrm{grad}_n\, T_0,$$

which defines the heat transfer intensity between the body and the surrounding medium.

A characteristic and obvious feature of the problem is that the origin from which the temperature is measured is unimportant. Actually, all the equations have the temperature only in positions where it is operated on by operators whose values do not change when the origin of the temperature measurement is assigned any constant value. This is explained by the fundamental physical properties of the process of heat redistribution, which is entirely due to the nonuniformity of the temperature field and is quite independent of the absolute level of the temperature. In fact, the system of analytical relationships being considered (in complete agreement with the physical nature of the process) is defined only in terms of temperature differences and not in terms of the temperature itself on any predetermined scale. It is therefore valid to fix some characteristic temperature, defined by the conditions of the problem, as the origin of temperature. In this case all differences take on the sense of positive or negative excesses of temperature.

Let

$$\vartheta \equiv T - T',$$

where ϑ is the temperature excess corresponding to the temperature T;

   T' is a temperature given by the conditions and taken as the zero reading.

The equations can be written as follows in terms of the temperature excess:

$$\frac{\partial \vartheta}{\partial t} = a \nabla^2 \vartheta \tag{3.2$'$}$$

and

$$\alpha \Delta \vartheta = \lambda \, | \, \mathrm{grad}_n \, \vartheta \, |_0. \tag{3.3$'$}$$

In its most general form, the solution to a problem stated in this way is of the form:

$$\frac{\vartheta}{\vartheta_0} = f\left(\frac{t}{t_0}, \frac{x_1}{l}, \frac{x_2}{l}, \frac{x_3}{l}; \ \pi_1, \ \pi_2, \dots; \ P_1, \ P_2, \dots\right). \tag{3.4}$$

The group of parametric groups P includes the ratios which define the geometric properties of the system. It can also contain ratios of various scale values of the temperature excess, as well as ratios of physical constants of the same nature $\left(\dfrac{a_2}{a_1}, \dfrac{a_3}{a_1}, \ldots ; \dfrac{\lambda_2}{\lambda_1}, \dfrac{\lambda_3}{\lambda_1} \right)$ which characterize the properties of different parts of the system, and ratios of the heat transfer coefficients at different parts of the surface (for example, at the inner and outer surfaces of a tube). As regards ratios of characteristic periods of time, it is only rarely that the conditions define two or more intervals of time.

Now let us consider the determination of the groups π. We will go through all the steps in detail, since it will be very helpful to follow the operations described earlier in general form even for a single actual case.

10. HOMOGENEOUS OPERATORS AND THEIR REDUCTION

Let us go back to the fundamental equation of the problem expressed in the form (3.2′):

$$\frac{\partial \vartheta}{\partial t} = a \nabla^2 \vartheta.$$

The following problem arises. The right-hand part of the equation contains the operator $\nabla^2 \vartheta \equiv \sum_1^3 \dfrac{\partial^2 \vartheta}{\partial x_k^2}$, which represents the sum of three terms. At first sight each of these terms should be considered as an independent operator in constructing the groups. In this case, we would obtain (e.g., after dividing by the derivative in the left-hand part of the equation) three relative operators, and hence three dimensionless groups. However, this result is very questionable, since it would be impossible to decide just what the specific properties of each of them would be.

The only difference which can be seen in these groups (if we consider them in the form in which they are obtained) is that they contain different dimensions: x_1, x_2 and x_3. However, these dimensions are only the projections of the single characteristic dimension l. As we have seen, this characteristic dimension must be used in all the groups by combining them with the ratio $\dfrac{x_{k_0}}{l}$. In this case all three groups become identical. Obviously our argument as to the sort of ratio must be improved. The fact is that the operator $\nabla^2\theta$ corresponds to a single group only, and so this operator must be considered as a whole in constructing the groups. Let us prove this.

Consider some three-dimensional temperature distribution, and let the temperature vary from a given value θ_1 to another given value θ_2 in a region of length l whose direction can be chosen quite arbitrarily (though, of course, it must be known). Using the general method, the lengths and temperatures are referred to the values at the beginning of the interval. (This means that the origins of the coordinate and temperature axes coincide with the initial point of the interval.)

In this case, as we move from the origin to the end of the given zone (i.e., along the radius-vector l) the temperature varies from zero to $\theta_0 \equiv \theta_2 - \theta_1$, and the coordinates vary from zero to x_{10}, x_{20}, x_{30}.

Now let us separate the distribution into three one-dimensional distributions along the coordinate axes. We can easily carry out the reduction operation for each of these distributions. Thus

$$\frac{\partial^2\theta}{\partial x_k^2} \rightarrow \frac{\theta_0}{x_{k_0}^2} \quad \text{or} \quad \frac{\partial^2\theta}{\partial x_k^2} = \frac{\partial^2\theta}{\partial X_k^2}\,\frac{\theta_0}{x_{k_0}^2}\,,$$

where, according to the convention established earlier, the capital letters denote relative values of the variables: $\theta \equiv \dfrac{\theta}{\theta_0}$ and $X_k \equiv \dfrac{x_k}{x_{k_0}}$.

Following the general rules for constructing the groups, we must introduce the characteristic dimension l. The last equations can then be rewritten as:

$$\frac{\partial^2 \theta}{\partial x_k^2} = \frac{\partial^2 \theta}{\partial X_k^2} \left(\frac{l}{x_{k_0}} \right)^2 \frac{\theta_0}{l^2} \quad \text{or} \quad \frac{\partial^2 \theta}{\partial x_k^2} = \frac{\partial^2 \theta}{\partial X_k^2} \frac{1}{\cos^2 \beta_k} \frac{\theta_0}{l^2} \; ,$$

where β_k is the angle between the radius-vector l and the corresponding coordinate axis.

Thus for the operator as a whole we have

$$\nabla^2 \theta = \frac{\theta_0}{l^2} \sum_{1}^{3} \frac{1}{\cos^2 \beta_k} \frac{\partial^2 \theta}{\partial X_k^2} \; .$$

It is obvious that the sum in the right-hand part of the equation is completely defined by the dimensionless temperature distribution along the given direction, which is specified by the value of the angle β_k, and hence plays the part of a multiplying factor N. In this case the reduced group is the expression $\frac{\theta_0}{l^2}$, which means that the reduction operation can be written in the form

$$\nabla^2 \theta \rightarrow \frac{\theta_0}{l^2} \; .$$

These arguments remain valid regardless of whether we are considering a steady-state or an unsteady-state temperature field. The only difference is that the unsteady-state distribution is a function of time as well as of the coordinates.

If the problem is set up so that the temperatures θ_1 and θ_2 at the boundaries of the region of space must be regarded as functions of time, so that $\theta_2 - \theta_1 \equiv \theta_0 = \theta_0(t)$, then in the final expression for the group, $\theta_0(t)$ is replaced by the constant value θ_{00} given in the conditions by dividing by the relative variable $\frac{\theta_0(t)}{\theta_{00}}$. In this way the result obtained is valid for all conditions. Thus we have shown that for determining the form of the group, the operator $\nabla^2 \vartheta$ must be regarded as a single whole, that is, as if it represented any one of its component terms. Of course, this result is in no way connected with the physical nature of the quantity on which the operator operates, and is quite general. Therefore, we can write quite generally:

$$\nabla^2 \longrightarrow \frac{1}{l^2} .$$

It is easy to see that this property of the operator ∇^2 is due to the fact that it combines expressions which are identical in structure in the form of a sum. Actually, the component terms are derivatives of the same order of some quantity with respect to the coordinates, i.e., with respect to the components of a length. Consequently, their only difference is that different components of the same quantity serve as the arguments; in essence, these are the independent variable. This difference cannot appear in the formation of the reduced group. As a result, the operator can be represented by any of its terms.

An operator possessing this quality will be termed homogeneous. Obviously this property of an operator will lead to the interesting result that the form of the corresponding group will be quite independent of whether the distribution is one-, two-, or three-dimensional. This result is quite reasonable physically. The operator as a whole is related to some physical effect, and in a medium with specified physical properties, this can depend only on the distribution of the quantity in space and on nothing else.

Let us consider one other example of a homogeneous operator. This time let us discuss the properties of the operator

$$\operatorname{div} \vec{w} \equiv \frac{\partial w_1}{\partial x} + \frac{\partial w_2}{\partial x_2} + \frac{\partial w_3}{\partial x_3} ,$$

which, in contrast to the operator ∇^2, is applied to vector quantities. Here $\vec{w}$ is a vector (velocity, for instance) whose components are w_1, w_2, w_3. This operator, known as the divergence, plays an important part in the theory of continuous media.

It is easy to see that the present case differs from that above, since the derivatives are not of exactly the same (scalar) quantity, but of different components of the same (vector) quantity. As a result, the terms appearing in the operator are derivatives of the components of one quantity with respect to components of another quantity. However, this additional complication, compared with the previous case, is quite unimportant, since it does not make itself felt in the reduction operation; this can be seen clearly from the preceding section. In this case, we can write at once

$$\operatorname{div} \vec{w} \rightarrow \frac{w}{l},$$

or, in general form,

$$\operatorname{div} \rightarrow \frac{1}{l}.$$

The following obvious fact must be noted. The formal criterion of a homogeneous operator is that it must always be given in the form of a sum of terms which differ only by their subscripts. Using this criterion it is very easy to establish in any given case whether an operator is homogeneous. This process will be widely used.

11. THE FOURIER NUMBER AND ITS PHYSICAL SIGNIFICANCE. HOMOCHRONICITY

Let us now return to Eq. (3.2′). We have shown that this equation must be treated as though it consisted of two terms. As a result, there is only one relative parameter, which can be written as

$$d = \frac{a \nabla^2 \vartheta}{\dfrac{\partial \vartheta}{\partial t}}.$$

Reduction can be carried out with no difficulties. We must only remember that on bringing the group to its final form it is necessary to introduce into it some single characteristic value of ϑ, although the problem conditions may define several different scales to which the temperature distributions in space and time could be referred. Let us write out in detail the whole reduction operation:

$$\frac{a \nabla^2 \vartheta}{\dfrac{\partial \vartheta}{\partial t}} \rightarrow \frac{a \vartheta_l / l^2}{\vartheta_t / t_0} = \frac{a t_0}{l^2} \frac{\vartheta_l}{\vartheta_t} \rightarrow \frac{a t_0}{l^2} \frac{\vartheta_0}{\vartheta_0} = \frac{a t_0}{l^2}.$$

Here ϑ_l and ϑ_t denote the reference scales, which are different for the temperature distributions in space and in time. These quantities are eliminated by multiplying the numerator and denominator by the ratios $\frac{\vartheta_l}{\vartheta_0}$ and $\frac{\vartheta_t}{\vartheta_0}$ (where ϑ_0 is a characteristic value of ϑ). If the values of ϑ_l and ϑ_t are given by the conditions (in addition to ϑ_0), the ratios $\frac{\vartheta_l}{\vartheta_0}$ and $\frac{\vartheta_t}{\vartheta_0}$ must be put into the solution as parametric groups.

Thus we have derived the dimensionless group corresponding to the fundamental equation, which is $\frac{a t_0}{l^2}$. This group is appropriately known as the Fourier number, and is denoted as

$$\frac{a t_0}{l^2} \equiv \mathrm{Fo}. \qquad (3.5)$$

The Fourier number expresses a definite relationship between the rate of change of the conditions in the surrounding medium and the rate of rearrangement of the temperature field within the body. The length of time t_0, which governs the time rate of change of the external conditions and which is given quite independently of the conditions within the system, is combined in the form of a ratio with another length of time which characterizes the development of the process within the system and which is completely specified by its properties and expressed in terms of the parameters l and a, but is quite independent of the regime parameters: α, $\Delta\vartheta$, etc. Every individual case (of course, in the generalized sense) corresponds to a definite value of this ratio. Put another way, this means that all the similar phenomena are characterized by the same value of Fo, and, as a result, equal values of Fo are a necessary prerequisite for the phenomena to be similar.

It is easy to show why this ratio plays such an important part. A change in the conditions in the surrounding medium is a primary cause of a rearrangement of the temperature field within a body. The characteristic period of time t_0 for this change governs the rate of the disturbance which penetrates into the system from outside and disrupts the temperature distribution set up in the body. The system reacts to this disturbance by means of changes which occur at a rate which depends on its natural properties in a definite way. This is expressed by the fact that the time required for these changes to occur is proportional to the square of the system dimension and inversely proportional to the thermal diffusivity of the medium. It will be quite obvious that the nature of the developing process must depend quite considerably on the relationship between these two specific time periods.

The investigation of all unsteady-state processes, regardless of their physical nature, leads to similarity groups giving the relationships between the rates of development of various effects which affect the course of the processes. These are sometimes

termed homochronicity groups, i.e., the group for timewise similarity. Groups of this type differ greatly in structure since the
physical mechanism of the process is expressed in them. Complicated processes may have two (or more, in principle) homoch-
ronous groups of different structures.

12. THE BIOT NUMBER: ITS PHYSICAL STRUCTURE AND REPRESENTATION. QUANTITATIVE EVALUATIONS BASED ON THE BIOT NUMBER. DEGENERATION OF GROUPS

The Fourier number is a necessary but not unique quantitative
feature of the generalized case. In other words, equality of their
Fourier numbers is a condition which is necessary but not sufficient to ensure that phenomena are similar. This is a result of
the fact that in addition to the fundamental equation, the conditions
of the problem contain additional equations which determine the
conditions at the boundaries of the system. The additional equations correspond to their own similarity groups, which provide the
second quantitative feature of the generalized case (the second
necessary requirement for similarity of the phenomena).

A characteristic feature of the problem being investigated is
that it is formulated with boundary conditions of the third type. This
means that in setting up the problem, the temperature conditions
can be regarded as known only in the surrounding medium. The
boundary conditions are unknown and are given only indirectly in
terms of the heat transfer process between the system and the
surrounding medium. The effect of the heat transfer process on
the formation of the temperature field [as defined by Eq. (3.3)] is
given in the following way. The transfer of heat from the fluid to
the solid body (or from the solid to the fluid) occurs as a result of
a temperature driving force. (To fix our ideas, let us assume that

the fluid is warmer than the solid. In principle nothing is changed
if the flow is in the other direction.) The flow of heat in the solid
is caused by nonuniformity of the temperature, of which the
temperature gradient is a measure. If we consider the conditions
directly at the interface, it becomes obvious that there must be a
close, single-valued relationship between the temperature gradient
in the solid and the temperature driving force from the fluid to the
solid. Actually, the quantity of heat $\alpha \Delta T df$ passing into the solid
through an element df of its surface area on the fluid side must
pass completely into the bulk of the material, since the quantity
of heat used for increasing the internal energy of the volume
element $df dn$, where n is the normal to the surface, is a quantity
of a smaller order of magnitude. However, if the temperature
gradient at the surface is $|\operatorname{grad} T|_0$, the quantity of heat passing
into the solid is given by $\lambda |\operatorname{grad}_n T|_0$. It is easy to see that Eq. (3.3)
is exactly satisfied when these two expressions are equal. Thus
the temperature gradient in the solid is related to the temperature
driving force by two parameters, one of which (α) characterizes
the intensity of the heat transfer process between the fluid and the
solid, while the other (λ) characterizes the intensity of the heat flow
process within the solid. This is expressed by the relationship

$$|\operatorname{grad}_n T|_0 = \frac{\alpha}{\lambda} \Delta T.$$

The factor $\frac{\alpha}{\lambda}$ is termed the *relative heat transfer coefficient,*
and is usually denoted by h, so that the last equation can be rewritten as

$$|\operatorname{grad}_n T|_0 = h \Delta T$$

or, in terms of the excess temperatures,

$$|\operatorname{grad}_n \vartheta|_0 = h \Delta \vartheta.$$

There is a group corresponding to this equation; from the physical significance of the equation, it must define in some specific way the nature of the relationship between the temperature conditions in the surrounding medium and the temperature distribution in the system. Let us derive this group. Omitting the very simple intermediate steps, it is easy to arrive at

$$\frac{\alpha l}{\lambda} \equiv hl.$$

(In addition, of course, we obtain temperature and geometric groups of the parametric type.)

This group is termed the Biot number:

$$\frac{\alpha l}{\lambda} \equiv \mathrm{Bi}. \tag{3.6}$$

Thus the group corresponding to the heat transfer equation and serving as an expression for boundary conditions of the third type is found to be the product of the relative heat transfer coefficient and a characteristic dimension. It unites a parameter characterizing the geometric properties of the system and two thermophysical parameters: one (α) from the fluid to the solid and the other (λ) within the solid. There is no doubt that these parameters have an important effect on the temperature conditions of the process, and therefore their appearance in the group is to be expected. However, it is still not clear just what properties of the physical conditions of the process are defined by the Biot number. Let us try to indicate the role of this group as a characteristic of the temperature conditions of the process.

Setting up the temperature conditions in the solid body and in the surrounding medium is complicated by the fact that these conditions are defined in quite different ways. The direct subject of the investigation is the processes occurring within the solid body. These are studied in full detail. The final solution corresponds to a sum of information sufficient to permit the temperature field within the body to be set up for any given moment of time. The basic mathematical form used in the analysis of the temperature conditions in the body—the temperature gradient—is an operator which makes it possible to set up a detailed and accurate physical picture of the situation.

In contrast, the transfer of heat in the surrounding medium plays a purely subsidiary part in this problem. It is of interest only to the extent necessary for studying the main process occurring in the solid body. In general, the temperature distribution in the medium is not considered, since it is sufficient to determine its overall effect in the form of the temperature driving force.*

Finally, it is necessary to set up some quantitative relationship between two fields, one of which is characterized by a gradient and the other by a driving force. Only homogeneous characteristics can be compared in the form of a ratio. It is impossible to derive the temperature gradient at a given point in the surrounding medium from a given value of the temperature driving force (this would mean that we could define the structure of the field in more detail without all the additional information). However, it is quite possible to make the reverse change, and give some temperature difference from the temperature gradient at each point. Let us carry out this operation for the temperature field of the solid body.

Mark off an interval of length l from the interface along the normal to the interface into the depth of the body, i.e., in the direction of the internal normal. The absolute value of the difference

---

*As we have seen, this is satisfactory even in the case of problems given with boundary conditions of the third type. Generally the surrounding medium need no longer be considered if the problem is formulated with boundary conditions of the first or second types.

between the temperatures at the ends of this section is termed the temperature drop over the length l, and is denoted as δT (or $\delta \vartheta$). Obviously the temperature drop is defined by

$$\delta T = \left| \int_0^l \frac{\partial T}{\partial n} \, dn \right| = \left| \frac{\overline{\partial T}}{\partial n} \right| l \equiv |\operatorname{grad}_n T| \, l,$$

or, in terms of the excess temperature,

$$\delta \vartheta = |\overline{\operatorname{grad}_n \vartheta}| \, l.$$

The bar above a quantity indicates an averaged value.

Here the distance l is taken to be a characteristic dimension of the body. In this case, the quantity δT acquires the significance of a temperature drop which is characteristic of the temperature field in the solid body. This characteristic temperature difference can be compared directly with the temperature driving force, i.e., with the temperature difference which characterizes the conditions in the surrounding medium; thus

$$\frac{\delta T}{\Delta T} = \frac{\alpha l}{\lambda} \frac{|\overline{\operatorname{grad}_n T}|}{|\operatorname{grad}_n T|_0}$$

or

$$\frac{\delta T}{\Delta T} = \operatorname{Bi} \frac{|\overline{\operatorname{grad}_n T}|}{|\operatorname{grad}_n T|_0}. \qquad (3.7)$$

Similarly, in terms of excess temperatures,

$$\frac{\delta \vartheta}{\Delta \vartheta} = \operatorname{Bi} \frac{|\overline{\operatorname{grad}_n \vartheta}|}{|\operatorname{grad}_n \vartheta|_0}. \qquad (3.7')$$

Now the role of the Biot number as a quantity establishing a special type of relationship between the temperature distributions in the body and in the surrounding medium is quite clear. Obviously the Bi number is *an approximate measure of the ratio between the temperature drop in the body and the temperature driving force* between the medium and the body. It is clear that this becomes more accurate the less the mean temperature gradient (over the length l) differs from the value at the boundary. In the case of smooth curves of comparatively small curvature, the mean and exact values of the derivatives do not differ by very much (they are quantities of the same order in all cases). However, if we consider a curve with a region of marked curvature, the values of the gradient at various points may be quite different from the mean value. This type of distribution is characteristic, for instance, of the initial period of rearrangement of a temperature field (theoretically speaking, the gradient at the surface tends to infinity at the initial moment of the process, while the values at all the other points differ little from zero). All arguments on the relative sizes of the temperature drop and temperature driving force based on the value of Bi must be used with appropriate care.

The possibility of determining the order of the ratio between the temperature drop and the driving force from the given value of Bi is of considerable importance in carrying out numerical solutions. The two limiting cases of very large and very small values of Bi are of particular interest in this connection.

In the first case (Bi $\gg$ 1) the temperature driving force is a very much smaller quantity than the temperature drop. This means that it can be neglected, and that the temperature of the body surface can be identified with the temperature of the surrounding medium. In this case the surface temperature is given directly by the conditions, and the problem is greatly simplified. Thus the case of very large Bi corresponds to a reversion of boundary conditions of the third type of those of the first type.

In the second case (Bi $\ll$ 1), on the other hand, it is possible to neglect the temperature drop in comparison with the temperature driving force, and hence it can be assumed that at any instant the temperature is uniform over the whole body. Of course, this also leads to a considerable simplification of the problem.

Estimation of the temperature conditions by means of the Biot number can be applied very successfully in the course of obtaining a solution. Naturally it is necessary to check on the validity of these estimates in all cases (i.e., it is necessary to show that the averaged value of the temperature gradient and its actual value at the surface of the solid are quantities of the same order).

It is useful to note a useful feature which is characteristic of both these cases. As we have explained, boundary conditions of the third type degenerate to those of the first type in the case of very large Bi. In essence, this means that the Biot number disappears from the solution (since the heat transfer equation disappears) and hence is eliminated from among the governing dimensionless groups. There is an analogous result in the second, opposite case (Bi $\ll$ 1). This arises from the fact that now the heat conduction process no longer has to be considered. It is very interesting that the Fo number is also obviously eliminated from the solution. It is not difficult to show that under these conditions the two groups are combined, forming a new group in the form of the product BiFo. Actually, if it can be assumed that at any instant during the process the temperature is uniform over the whole volume, then the fundamental equation can be written as

$$c\rho V d\vartheta = \alpha\vartheta f dt.$$

According to this

$$d = \frac{\alpha\vartheta f dt}{c\rho V d\vartheta} \rightarrow \frac{\alpha t_0}{c\rho l} = \frac{\alpha l}{\lambda} \frac{\lambda}{c\rho} \frac{t_0}{l^2} \equiv \mathrm{BiFo}.$$

Thus in both limiting cases (Bi $\gg$ 1 and Bi $\ll$ 1) the group Bi is eliminated: in the first case it simply drops out, and in the second case it is combined with Fo. It is not difficult to demonstrate the physical significance of this fact. Any group represents an approximate measure of the relative intensity of two physical effects. If the condition is also added that the effects being compared are incommensurate, the group involved begins to increase or decrease without limit, and loses its significance. This is the case that we have encountered here. The condition $Bi \gg 1$ means that the intensity of the heat transfer process is immensely large compared with that of the heat transfer process within the body. On the contrary, the condition $Bi \ll 1$ corresponds to an enormously large intensity of the heat conduction process. If any group assumes a very large or a very small value, and as a result of this disappears, we say that it degenerates. (Later we will meet several different cases of group degeneration and consider them in more detail.) Here we need only note that the disappearance of a parametric group from the conditions of the problem is an elementary form of degeneration, since it causes "incommensurability" (for example, the case of an infinitely long cylinder is $\frac{l}{R} \gg 1$, and that of a flat disc is $\frac{l}{R} \ll 1$, and in both cases $\frac{l}{R}$ cannot appear in the solution).

Our discussion of the properties of the Biot number can be given an interesting and useful graphical interpretation. In Fig. 2, let n be the direction of the normal to an element of the solid surface which is in contact with the surrounding medium. The graph is drawn in terms of the coordinates (ϑ, n), with the origin at the base of the normal. If the points A and B on the ordinate axis correspond to the temperatures of the medium ϑ_m and of the solid ϑ_b, the length AB determines the temperature driving force. Also, suppose that the temperature distribution in the body starting at the boundary point B is given by the dashed curve shown in the figure. By drawing the tangent to this curve through the point

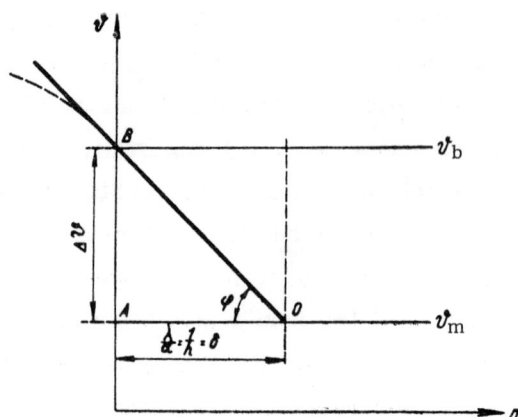

Fig. 2

B, we obtain an angle φ which satisfies the obvious condition $\tan \varphi = |\mathrm{grad}_n \vartheta|_0$, and hence $\tan \varphi = h\Delta\vartheta$ (it must be remembered that we always consider the absolute values of the quantities). On the other hand, $\tan \varphi = \dfrac{AB}{AO}$, where AB, as already mentioned, represents the temperature driving force $\Delta\vartheta$. By comparing the two expressions for $\tan \varphi$, we find directly that $OA = \dfrac{1}{h} = \dfrac{\lambda}{\alpha}$. Thus a quantity which is the inverse of the relative heat transfer coefficient can be defined as a length. In itself, this result is not of particular interest, since it can be derived directly from elementary arguments about the dimensions of the quantities [for example, from the requirement that both parts of Eq. (3.3) must have the same dimensions]. However, there is an important property of this length. If we draw a straight line in terms of these same coordinates which is parallel to the ordinate axis and at a distance $\dfrac{\lambda}{\alpha}$ from it, the intersection of this line with the horizontal $\vartheta = \vartheta_m$ is given by the point O, which has the following important property: regardless of the details of the process, the tangent to the temperature distribution curve drawn through its final point (i.e., through its point on the surface of the solid body) must pass through the point O. In other words, we can say that the point O is a pole through which we can draw a family of lines which are the tangents to the temperature curves at the points where they intersect the surface. The point O is therefore called the director point.

The picture of the process represented in the figure corresponds to some instantaneous state of the body and surrounding medium. This changes continuously, since the elements vary as a function of time. It is very important that the length $\frac{\lambda}{\alpha}$ remains constant throughout all these changes, and hence the position of the vertical line drawn through the point O remains completely fixed. As the temperature of the surrounding medium ϑ_m changes, the point O moves, but without leaving the vertical. For instance, if ϑ_m varies periodically, the point O will perform oscillations along the vertical line. Thus a straight line parallel to the ordinate axis and at a distance $\frac{\lambda}{\alpha}$ from it is the geometric locus of the director point.*

On passing from one value of Bi to another, the effect is obviously represented on the figure by a shift of the director point relative to the ordinate axis, and a corresponding change in the slope of the tangent. As Bi increases to infinity, the director point approaches the axis. The length of the tangent between the director point and the axis decreases continuously, and finally it degenerates to a point, which is obviously the initial point on the temperature

---

*Of course, this result depends importantly on the assumption that λ and α are constants. If the intensity of the heat transfer process changes with time, the director point describes a trajectory whose equation can be given from the conditions [for example, in parametric form, $\vartheta_m = \vartheta_m(t)$ and $\alpha = \alpha(t)$].

curve of the solid body. At the same time the distance along the axis between the values ϑ_m and ϑ_b also contracts to a point. As Bi tends to zero, the director point moves away to infinity, and the tangent tends to become horizontal $\left(\dfrac{\partial \vartheta}{\partial n} = 0\right)$.

If the temperature of the surrounding medium Bi remains unchanged throughout the whole process, so that the rearrangement of the temperature field is caused simply by changes in the conditions in the surrounding medium (at the initial moment), the position of O is completely fixed on the graph by the intersection of the vertical line $\dfrac{1}{h} = \dfrac{\lambda}{a} =$ const and the horizontal line $\vartheta = \vartheta_m =$ const. As a result, only one director point is obtained for the whole process. This serves as a pole from which originates a whole family of straight lines which are the tangents to the temperature curves at the surface of the solid at different instants during the process. This special case of heating (or cooling) a solid in a medium of constant temperature and at a constant heat transfer intensity is of great interest. We will consider it in detail a little later.

Our idea of representing the quantity $\dfrac{\lambda}{a}$ by means of distance which is characteristic of the process can be developed somewhat further. Let us denote the distance from the ordinate axis to the director point by δ $\left(\text{so that } \dfrac{\lambda}{a} \equiv \delta \right)$. In this case the Biot number can be written in the form

$$\text{Bi} = \frac{l}{\delta} . \qquad (3.8)$$

By writing it in this way, we arrive at a new aspect of the Bi number—it is a ratio of two characteristic lengths. One of these, l, is given directly by the conditions as the dimension which defines the geometric properties of the system; the other, δ, does not appear as a length in the conditions. Instead, it is defined implicitly by the parameters λ and α, which characterize the thermal conditions of the process. The combination of these two parameters into a single one means that the two parts of the process corresponding to Eq. (3.3) do not require both parameters to be introduced separately: in studying them it is sufficient to know the relative value of these parameters. In any case, this was already clear from the fact that Eq. (3.3) led to the formation of the relative heat transfer coefficient h.

We now have the new state of affairs that it is possible to reduce the definition of the physical conditions at the surface of the solid to a single characteristic length. Comparison of this length with the characteristic dimension of the body leads to a new form of the Biot number. Let us discuss the physical significance of this comparison.

Imagine that the region in Fig. 2 between the ordinate axis and the vertical line at a distance δ from it corresponds to a layer of the solid material instead of to the liquid surrounding the solid body. The straight line BO will be regarded as the graph of the temperature distribution in this layer (i.e., as an extension of the temperature curve of the main body). In contrast to the dashed curve, whose shape depends on the real conditions of the process, the temperature profile in the additional layer is given by a straight line. It is not difficult to describe the physical circumstances under which this imaginary distribution could occur. It is known that the temperature changes linearly over the thickness of a uniform thin plate (i.e., a plate whose thickness is small compared with its two other dimensions) through which a steady-state heat flux passes (the thermal conductivity is taken to be a constant quantity). Disregarding the exact relationship to the real process, let us imagine that a flat plate adjoins the body being investigated, and that a steady heat flux passes through it. It is easy to determine the density of this heat flux if the thermal conductivity of the plate is known. Suppose that the plate is made of the same material as the main solid body. Then

$$q = \frac{\lambda}{\delta}(\vartheta_m - \vartheta_b).$$

But, by definition, $\frac{\lambda}{\delta} = \alpha$. Hence

$$q = \alpha(\vartheta_m - \vartheta_b) = \alpha \Delta \vartheta.$$

Thus the imaginary heat flux is identical to the real flux occurring at a particular instant in the process (i.e., to the flux at the particular instant which is actually occurring in the process which is being represented).

This result allows us to give a definition of the quantity δ which is based on clear and simple physical concepts. Obviously δ can be defined as the thickness which a plate made of the same material as the main solid body must have in order that a steady heat flux $\alpha \Delta \vartheta$ passing through it should lead to a temperature drop $\Delta \vartheta$ (i.e., to a temperature drop equal to the real temperature driving force). In the Bi number this quantity is compared ratiowise with the characteristic dimension of the body l; over this distance l normal to the surface the temperature changes by $\delta \vartheta$. If the temperature also changed linearly within the body, the ratio $\frac{l}{\delta}$ (i.e., Bi) would be exactly equal to $\frac{\delta \vartheta}{\Delta \vartheta}$. However, the real temperature distribution is curvilinear, and so the Bi number is only an approximate measure of the ratio $\frac{\delta \vartheta}{\Delta \vartheta}$, as we showed earlier.

13. MONOTONICALLY VARYING TEMPERATURE FIELDS. THE FOURIER NUMBER AS A DIMENSIONLESS FORM OF TIME. ONE-DIMENSIONAL FIELDS

We have shown that in the general case the problem is char-

acterized by two groups: $\mathrm{Fo} \equiv \dfrac{at_0}{l^2}$ and $\mathrm{Bi} = \dfrac{al}{\lambda}$. Thus Eq. (3.4)

assumes the form

$$\frac{\vartheta}{\vartheta_0} = f\left(\frac{t}{t_0}, \; \frac{x_1}{l}, \; \frac{x_2}{l}, \; \frac{x_3}{l}; \; \mathrm{Fo, \; Bi;} \; P_1, \; P_2, \; \ldots \right) .$$

Both groups possess the interesting property that they do not involve the temperature, though it is quite clear that they characterize the properties of a temperature field. However, this is by no means a contradictory state of affairs. The reason for this feature must be sought in the structures of Eqs. (3.2) and (3.3), from which the groups were obtained.

It is easy to see that *both equations are linear and homogeneous with respect to temperatures*. This property expresses itself directly in the fact that the reduced groups which we obtain for the relative operators contain the temperature to the same power in the numerators and denominators (as a result of which they cancel out and cannot appear in the groups). From a physical point of view this means that the relative intensities of the effects which are included in Fo and Bi are independent of the temperature scale value ϑ_0, since a change in this quantity has exactly the same influence on the intensities of any pair of the effects. (Or, put another way, if we consider a group of similar phenomena as the same phenomenon given in terms of different scales, we are quite unable to associate the scale of the temperature field with the scales of the other physical fields.) Thus, temperature can appear among the arguments of Eq. (3.4) only in the form of parametric groups.

Now let us turn our attention to the following important fact. Eq. (3.4$'$) corresponds to the very general case when the physical conditions in the surrounding medium do not remain constant. Under these conditions the manner in which they change must be specified (for example, in the practically important case of a fully established periodic process—a steady-state periodic process—the period and amplitude of the temperature changes must be given). In this way some time period t_0 is defined which can be used as the scale value for time. In addition, at least one temperature group of the parametric type must be introduced among the arguments in the form of the ratio between the amplitude of the time-wise temperature change and the initial temperature driving force.

Equation (3.4$'$) has been written in the typical form of a generalized equation. However, it is not always possible to proceed further with the solution of a temperature-field problem in this form. Very often we have to consider processes of monotonic temperature changes in a solid body in the presence of constant external conditions. In this case the problem conditions do not give any characteristic time interval, and it is therefore impossible either to construct the group $\frac{at_0}{l^2}$ or to express the relative time in the form $\frac{t}{t_0}$. Thus the familiar situation arises which we discussed earlier in connection with the problem of combining groups with relative variables.

Following the general method, it is necessary to find some combination of the group $\frac{at_0}{l^2}$ and the relative variable $\frac{t}{t_0}$ that does not contain the parameter t_0, which is not given in the conditions. Obviously this is done by forming the product of the two quantities. Thus

$$\frac{at_0}{l^2} \cdot \frac{t}{t_0} = \frac{at}{l^2} \ .$$

This expression is a typical example of a dimensionless variable. The dimensionless time is obtained by referring the varying
value of the time t not to its parametric value t_0 (since we are not
provided with this value) but to the group of quantities $\dfrac{l^2}{a}$ which is
equivalent to this value.

The actual significance of this equivalence of a parametric
value of a given quantity and the group of other quantities sub-
stituted for it has been considered earlier in the general discussion.

In the present specific case we can go somewhat further. We
will attempt to explain the physical importance of the quantity $\dfrac{l^2}{a}$
as a time interval characteristic of the process being studied. In
this connection it must be noted that the analytical expression for
the excess temperature, measured with respect to the constant
temperature of the surrounding medium as the zero value, contains
a factor of the form $e^{-f(Bi)\frac{at}{l^2}}$. This factor is an expression of the
law giving the timewise decrease of the temperature field nonuniformities. Its structure is in complete agreement with the
general conclusions obtained earlier by using the method of
generalized variables. It is quite obvious that the rate of change
of the temperature with time must depend on the heat transfer
intensity between the body and the surrounding medium. This
effect is expressed by $f(Bi)$. For given conditions at the surface
(specified value of Bi), the factor assumes the form $e^{-k\frac{at}{l^2}}$, where
k is a dimensionless constant (some number). The significance of
this result is clear. If a definite regime is established for the

interaction of the body and the surrounding medium, the temperature field arising as a result of this interaction will be formed at a rate which depends on the natural properties (geometric and physical) of the body.

The degree of attenuation of the temperature nonuniformities (which exist in the body at all intermediate moments of the process and disappear only in the ultimate equilibrium state) is determined not by the time which has elapsed since the process began, but by its ratio to the quantity $\frac{l^2}{a}$. Thus the role of this quantity as a characteristic time period is quite obvious. Let us denote it as τ.

The factor of interest can then be written as $e^{-k\frac{t}{\tau}}$. If it is assumed that $t = \tau$, this becomes e^{-k}. Thus the characteristic time period has the significance of the period of time during which the temperature nonuniformities decrease by e^k times.

Of course it would be quite incorrect to regard the group $\frac{at}{l^2}$ as a similarity group. As long as we are discussing $\frac{at_0}{l^2}$ we are dealing with an expression which is fixed by the definite conditions for the development of the process. It is only phenomena which possess the same values of these conditions (i.e., which have the same value of $\frac{at_0}{l^2}$) which can be similar to one another. As a result, this quantity can be taken to be a similarity group without further discussion.

On replacing the parameter t_0 by the variable value t the situation changes radically. Generally speaking, the quantity $\frac{at}{l^2}$ does not express any preliminary requirement in the circumstances of the process, and the fact that it is equal for different phenomena is emphatically not a prerequisite for similarity. If all the other conditions for similarity are satisfied (equality of the values of Bi and of the parametric criteria, similar initial and surface fields), the rearrangement processes of the temperature fields and all the accompanying processes will be completely similar. Clearly this similarity occurs only as a result of their exact coincidence. The group $\frac{at}{l^2}$ also gives a valid comparison of processes with respect to time: fields which have the same value of the group are similar to one another.

If we figuratively represent the processes to be compared as infinite sets of mutually interchangeable temperature fields, it is safe to say that there will be an analog in each of the remaining sets for any field chosen at random from a given set; this analog will be in the form of an exactly similar field. The condition that the values of $\frac{at}{l^2}$ should be equal provides a method for finding these similar fields. Exact equality of the values of the Fourier number $\frac{at_0}{l^2}$ is a requirement for similarity; exact equality of the values of the Fourier number $\frac{at}{l^2}$ is the principle for selecting corresponding moments of time (i.e., the moments when the fields are similar). In relative form, this difference can be expressed as follows: the *first Fourier number is a quantitative characteristic of a generalized case; the second is a relative form of the time variable.*

Thus, processes involving monotonic changes in the temperature of a body (heating, cooling) in a medium of constant temperature are characterized by only one similarity group, $\mathrm{Bi} \equiv \frac{al}{\lambda} \equiv hl$. Depending on the exact way the problem is stated, it will also be necessary to introduce various groups of the parametric type to define the geometric properties of the system and the features of the temperature distributions given by the conditions.

Frequently technical applications make use of calculated results referring to problems which are so simple that in general it is unnecessary to have any parametric group at all. Such problems arise in the study of processes which occur in thin plates, infinitely long cylinders, and spheres, when the solid body is held for a sufficiently long time in a medium at a given constant temperature, and is then transferred to a medium with a different, but also constant, temperature (i.e., the process involves the rearrangement of a uniform temperature field in these simple solids into a uniform field of another temperature). Under these conditions it is unnecessary to introduce any groups of the parametric type, since we are given only one dimension (the thickness of the plate, diameter of the cylinder or sphere) and only one excess temperature (the initial temperature of the body).

The characteristic feature of these problems is that they are *one-dimensional*, i.e., they are concerned with the investigation of temperature distributions which are functions of only one co-ordinate (the temperature distributions along the normal to the plate surface, or along the radius of the cylinder or sphere). As a result, only one space coordinate appears among the arguments. Obviously the same coordinate defines the direction of the heat flux.

Thus in this case the group of arguments is limited to three quantities: one parameter and two independent variables. Equation (3.4) can be greatly simplified and reduced to the form

$$\frac{\vartheta}{\vartheta_0} = F\left(\text{Bi}; \frac{at}{l^2}, \frac{x}{l}\right). \tag{3.9}$$

Here $\frac{x}{l}$ is the relative value of the coordinate along which the temperature change occurs (the distance from the center of the plane expressed as a fraction of the half-thickness of the plate; the distance from the axis of the cylinder or from the center of the sphere as a fraction of the radius). By fixing this value, we obtain a solution for some geometric locus of a point of specified temperature: this may be a plane normal to the heat flux (for instance, the value $\frac{x}{l} = 0$ corresponds to the central plane, and the value $\frac{x}{l} = 1$ corresponds to the boundary plane); it may be a coaxial cylindrical surface (with $\frac{x}{l} = 0$, this is the cylinder axis, and with $\frac{x}{l} = 1$ it is the outer surface); or it may be a concentric spherical surface (with $\frac{x}{l} = 0$, this is the center of the sphere, and with $\frac{x}{l} = 1$ it is its outer surface). In this way, the relative temperature $\frac{\vartheta}{\vartheta_0}$ is given as a function of the relative time $\frac{at}{l^2}$ with Bi as a parameter.

This form of presentation of the results of the solution (in tables or by means of graphs) is very widely used. However, it is convenient to change it somewhat. The characteristic dimension occurs in the generalized variable $\frac{at}{l^2}$ as well as in the parameter $\mathrm{Bi} = hl$. This makes it somewhat more difficult to determine the separate effects of the natural dimensions of the system and those of the time. It is therefore helpful to eliminate the quantity l from the expression for the dimensionless time. Thus

$$\frac{at}{l^2} (hl)^2 = ah^2 t.$$

The deficiency mentioned above has now been removed: when the thermal conditions of the process are given (fixed values of λ, a and α) the parameter is proportional to the dimension l, and the independent variable is the time t.

14. STEADY-STATE TEMPERATURE DISTRIBUTIONS IN PLATES. THERMAL CONDUCTIVITY AND THERMAL RESISTANCE. NEW ASPECTS OF THE BIOT NUMBER

Up to now we have considered the problem of monotonically changing temperature fields with the assumption that the whole surface of the solid body is in contact with a fluid with a constant temperature. Under these conditions, the final steady state corresponds to a uniform temperature field: all points of the body reach the same temperature, which is equal to that of the external medium. Now let us assume that the limiting planes of a plate are in contact with fluids of different temperatures. In this case (which is of great practical importance) the steady-state temperature field of the plate will not be uniform. The temperature differences cannot disappear, since they are maintained by the existence of the temperature difference between the fluids, which leads to the transfer of heat from the warmer to the cooler fluid through the plate separating them. Obviously, under these conditions the steady state can only be reached when the heat supplied to any arbitrarily thin layer of the plate is equal to the heat lost by this layer. It is easy to see that this specific equilibrium condition, which can be reduced to the requirement that $\left(\lambda \dfrac{d\theta}{dx} \right)_x = \left(\lambda \dfrac{d\theta}{dx} \right)_{x+dx}$ at any value of x, can be satisfied by only one temperature distribution, which will depend on the manner in which the thermal conductivity λ varies with temperature. If it is assumed that λ does not vary, the equilibrium condition becomes

$$\frac{d\theta}{dx} = \text{const.}$$

In the steady state, therefore, the temperature of the plate varies linearly along the coordinate which coincides with the direction of the heat flux and is normal to the main surfaces of the plate. This is a very simple case presenting no real difficulty for study, but it merits closer inspection because its *fictive distribution* (which leads to the constant value of the derivative, see p. 25) *is identical with the real distribution.* As a result of this interesting feature, the approximate evaluations given in terms of the groups can be used as exact quantitative results. Hence, under these conditions, $\mathrm{Bi} = \dfrac{\delta\vartheta}{\Delta\vartheta}$. In this case,

$$\frac{\delta\vartheta}{\Delta_1\vartheta} = \frac{\alpha_1 l}{\lambda} = \frac{l}{\delta_1} \text{ or } \frac{\delta\vartheta}{\Delta_2\vartheta} = \frac{\alpha_2 l}{\lambda} = \frac{l}{\delta_2},$$

whence

$$\frac{\Delta_1\vartheta}{\delta_1} = \frac{\delta\vartheta}{l} = \frac{\Delta_2\vartheta}{\delta_2}.$$

or

$$\Delta_1\vartheta : \delta\vartheta : \Delta_2\vartheta = \delta_1 : l : \delta_2.$$

On the other hand,

$$\Delta_1\vartheta + \delta\vartheta + \Delta_2\vartheta = \vartheta_1 - \vartheta_2 \equiv \vartheta_0,$$

where ϑ_0 is a characteristic temperature difference given by the conditions.

Thus the characteristic temperature difference is distributed among the two temperature driving forces and the temperature drop in proportion to the respective characteristic lengths. Using the notation $\delta_1 + l + \delta_2 \equiv L$, we have

$$\frac{\Delta_1\vartheta}{\vartheta_0} = \frac{\delta_1}{L}, \quad \frac{\delta\vartheta}{\vartheta_0} = \frac{l}{L}, \quad \frac{\Delta_2\vartheta}{\vartheta_0} = \frac{\delta_2}{L}$$

or

$$\frac{\vartheta_0}{L} = \frac{\Delta_1\vartheta}{\delta_1} = \frac{\delta\vartheta}{l} = \frac{\Delta_2\vartheta}{\delta_2}.$$

Figure 3 illustrates these results graphically. In setting up the system it is assumed for greater generality that the conditions at

the surfaces of the plate differ in the heat transfer intensities
(α_1 and α_2) as well as in the fluid temperatures (ϑ_1 and ϑ_2). It is to
be noted that it is convenient to take the (constant!) temperature
ϑ_2 of the cooler fluid as the reference temperature, and to use the
line $\vartheta_2 = \text{const}$ as the abscissa.

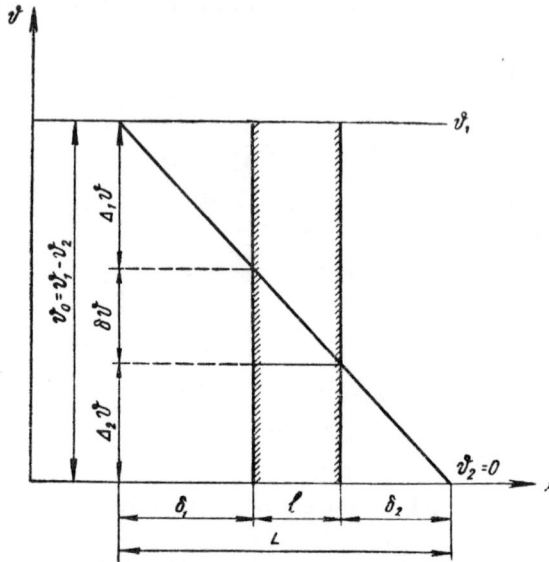

Fig. 3

This case is also of interest to us because its consideration
leads naturally to one of the forms of representing the group Bi.
Equation (3.7) defines this group as the ratio between two quantities
of the same type. This method of representation, when the group
formally (but only formally) assumes the form characteristic of a
group of the parametric type, is very useful, since then the role of
the group as a measure of the relative intensities of the particular
physical effects becomes quite clear. The group Bi is expressed
as the ratio of two characteristic lengths in Eq. (3.7). Let us now
show that it can be represented as a ratio of characteristic quantities of other physical types.

Let us still confine our attention to a steady-state process. By considering the successive transfer of heat from the fluid to the plate (from the plate to the fluid) and across the plate, we arrive at the following obvious double equation for the heat flux:

$$q = \alpha \Delta \vartheta = \frac{\lambda}{l} \, \delta \vartheta.$$

The structure of this expression suggests that the factors $\frac{\lambda}{l}$ and $\alpha = \frac{\lambda}{\delta}$ can be regarded as *thermal conductances* (and the reciprocal quantities $\frac{l}{\lambda}$ and $\frac{1}{\alpha} = \frac{\delta}{\lambda}$ as *thermal resistances*) corresponding to the links in the chain of the constant heat flux. Thus it is now possible to represent Bi as the ratio of the thermal conductances (or thermal resistances) characteristic of the process. This method of representation stresses the role of Bi as the measure of the relative intensities of the heat transfer processes to the surrounding medium and to the plate itself at the plate surface.

The relationship illustrated in Fig. 3 assumes the following significance on the basis of the thermal resistance concept. The total resistance of the whole chain of processes $\frac{L}{\lambda} \equiv R$ is made up of a summation of three successive resistances:

$$R \equiv \frac{L}{\lambda} = \frac{\delta_1}{\lambda} + \frac{l}{\lambda} + \frac{\delta_2}{\lambda} \, .$$

The total temperature difference for the chain as a whole, ϑ_0, is distributed among its three elements in proportion to their thermal resistances. The expression for the heat flux density becomes $q = \frac{\vartheta_0}{R}$, or $q = k\vartheta_0$, where

$$k = \frac{1}{R} = \frac{1}{\dfrac{1}{a_1} + \dfrac{l}{\lambda} + \dfrac{1}{a_2}}$$

is the *heat transport coefficient*.

15. THE TEMPERATURE FIELD OF A BODY OF INFINITE DIMENSIONS. COMBINATION OF INDEPENDENT VARIABLES

Although they have been quite different in form, all the problems considered so far have been similar in the sense that the conditions have defined at least one characteristic dimension. This arrangement of the problem is typical of a very wide circle of quite different sorts of problems: its frequent occurrence, in fact, merely indicates how important the geometric properties of the system are for processes. Generally speaking, it seems obvious that the course of a process must depend directly on the geometry of the system. However, if the conditions are such that the extent of the system can be regarded as immeasurably large compared with the part occupied by the process, the effect of the geometric properties of the system is no longer important, and they generally disappear from the problem. In this case, the conditions contain no dimension which can be used as a reference scale. We have a very characteristic example of this in the problem of the temperature field in a body of infinite size.

Since it is impossible to introduce a parametric value of the length, the specific feature of this case consists in replacing the characteristic dimension l by the equivalent ratio $\frac{\lambda}{a} \equiv \delta$. Obviously this ultimately entails the following change in the solution: the time is represented in the same dimensionless form obtained earlier, $ath^2 \left(\equiv \frac{at}{\delta^2} \right)$, and the coordinate becomes a dimensionless variable of the type $\frac{ax}{l} \left(\equiv hx \equiv \frac{x}{\delta} \right)$, while Bi usually disappears. It is important to note that this form of solution is possible only when the problem is set up with boundary conditions of the third type.

In the investigation of processes occurring in infinite masses it often appears to be possible to give some value of the temperature directly (i.e., to consider the problem with first-type boundary conditions). Under these conditions, the heat transfer coefficient (and hence the Biot number) is usually eliminated. As a result, the only operations which can be used for eliminating the dimension l from the variables $\frac{x}{l}$ and $\frac{at}{l^2}$ are limited by the structures of the various combinations of these variables. It is not difficult to see that the only suitable combination results from combining them in the form of the groups $\frac{at}{x^2}$ or $\frac{x}{\sqrt{at}}$. In this way the time is reduced to dimensionless form by referring it to the characteristic time period $\frac{x^2}{a}$, and the coordinate is reduced to dimensionless form by referring it to the distance $\sqrt{at}$.

Here it is worth turning our attention to an unusual feature of this form of constructing the characteristic quantities, which are defined in terms of variable (rather than parametric) values, and are therefore variable quantities. The most important fact is that the independent variables are combined into one dimensionless group. They can be separated, since one or the other (in definite combinations with the physical constant a) can be used as the reference scale.

A group including variables is the only argument of the solution. This is a special feature of processes which occur in infinite masses. With this type of solution it is impossible to isolate the effects of each of the independent variables directly. Nevertheless, it is possible to draw certain conclusions about the development of the process. For instance, it is obvious that the depth to which a given temperature front penetrates is proportional to the square root of the time period and to a factor $\sqrt{a}$. It is also obvious that the time required for a wave front to penetrate to a given depth is proportional to the square of the depth and inversely proportional to a.

Comparing the results obtained for the problems considered above, it can be seen that the solutions are quire different structurally. This leads to the conclusion that the same fundamental equation can have different types of solutions, so that the fundamental equation itself does not determine the structure of the generalized variables in terms of which the solution must be given. The form of solution changes appreciably depending on the information expressed by the boundary conditions.

To sum up, it can be said that the form of the solution depends not only on the mechanism of the process being studied, but also on the specific physical circumstances defined by the statement of the problem.

16. THE TEMPERATURE FIELD IN A BODY WITH A DISTRIBUTED HEAT SOURCE. DIMENSIONLESS GROUPS INVOLVING TEMPERATURE. THE CASE OF SOLUTIONS WHICH ARE COMPLETE EXCEPT FOR A NUMERICAL CONSTANT

In conclusion, let us review some of the discussions which are of particular interest as regards the problem of temperature fields in solid bodies. We have already had occasion to note that we do not obtain groups containing the temperature in the study of temperature fields; it was explained that this important feature of this type of problem was a result of the linear and homogeneous nature of all the equations being considered with respect to temperature. However, there is a very wide class of problems, which are of very great interest both theoretically and by reason of their applications, in which the characteristic fundamental equations are not homogeneous. The equations defining the process are not homogeneous with respect to temperature in all cases when there are effects at the boundary or within the system which are connected with the liberation or absorption of heat. In the theory of temperature fields these effects (which are of various physical natures) are introduced into the problem in the form of heat sources, which may be positive or negative (sinks), and which may be distributed continuously or localized in some way in space. Depending on the intensity of the effect being considered, the sources are characterized quantitatively by their strengths (specific outputs), i.e., by the quantities of heat involved per unit time per unit volume. The source strength may be given directly or calculated from the equations which define the intensity of the effect being studied. It is not expressed in terms of a temperature change of the body. Thus, when this quantity appears in the equation, the property of homogeneity is inevitably lost. We will try to explain what this implies by means of a simple example.

Let us consider the very simple problem of the steady-state temperature field in a homogeneous body with a uniformly distributed heat source of constant strength W; the body is surrounded by a medium at constant temperature. In this case the fundamental equation of the problem is

$$\nabla^2 \vartheta + \frac{1}{\lambda} W = 0. \tag{3.10}$$

The initial condition obviously disappears, since the problem is steady-state. The conditions at the boundaries of the system are defined as previously by boundary conditions of the third type:

$$\alpha \vartheta_m = \lambda \, | \, \mathrm{grad}_n \, \vartheta \, |_0.$$

Here the excess temperature ϑ is calculated with respect to the temperature of the surrounding medium.

On passing to the analysis of the equation, it is noted immediately that the statement of the problem gives us no characteristic temperature value. Hence only one similarity group is obtained, $\mathrm{Bi} = \frac{ad}{\lambda}$. The group $\frac{\vartheta\lambda}{l^2 W}$, which corresponds to Eq. (3.10), obviously has the significance of a dimensionless varying temperature, and is not a similarity group. A special feature of this expression is that it compares the temperature excess with quantities of other physical types, which together form the characteristic temperature $\frac{l^2 W}{\lambda}$. This is the first time that we have encountered this type of ratio. Physically it arises because the temperature field is set up and maintained in the steady state by the action of the heat source. This characteristic temperature cannot be formulated in the absence of a heat source, in which case the characteristic temperature must be given directly by the conditions.*

Thus, in terms of generalized quantities, the solution assumes the form

$$\frac{\vartheta}{\dfrac{l^2 W}{\lambda}} = F\left(\mathrm{Bi};\ P_1,\ P_2, \ldots;\ \frac{x_1}{l},\ \frac{x_2}{l},\ \frac{x_3}{l}\right), \qquad (3.11)$$

where $P_1,\ P_2, \ldots,$ are geometric criteria of the parametric type.

In the case of a one-dimensional problem (for instance, the temperature field in a thin plate) the equation assumes the simpler form

$$\frac{\vartheta}{\dfrac{l^2 W}{\lambda}} = f\left(\mathrm{Bi};\ \frac{x}{l}\right). \qquad (3.11')$$

---

*As we have shown, the conditions do not define a characteristic temperature value under the conditions of a steady-state problem. However, this fact cannot lead to any vagueness. A steady temperature field in a body which interacts over its whole surface with a medium having the same temperature can only arise in the presence of a heat source. In the absence of a source, the whole body arrives at a temperature equal to that of the surrounding medium, and the problem loses its sense. A steady-state temperature field in the absence of heat sources can occur only if the surface of the body is in contact with at least two fluids at different temperatures. In this case the difference between these temperatures serves as the reference scale.

The solution is quite different if the problem is set up with boundary conditions of the first type (for example, in the case of a very high heat transfer intensity, with moderate thermal conductivity and with the dimensions of the body not too small). In this case, the temperature at the surface of the body is known directly, and the problem reduces to establishing the excess temperature distribution (the excess temperature here is measured with respect to the given temperature at the surface). Using the same symbol ϑ as before for the excess temperature, we have:

$$\frac{\vartheta}{\frac{l^2 W}{\lambda}} = \varphi\left(\frac{x}{l}\right). \tag{3.12}$$

Because of the fundamental limitations of our method of analysis, we are not, of course, in a position to determine the form of the function φ. However, we can state that its value must lie in the range from $\varphi_{max} = \varphi(0)$ at the median plane of the plate to $\varphi_{min} = \varphi(\pm 1)$ at the plate surfaces. Obviously the value of φ_{min} is zero. The maximum value of the function is unknown. However, $\varphi(0)$ certainly represents some constant number. Hence at the median plane of the plate

$$\frac{\vartheta}{\frac{l^2 W}{\lambda}} = \text{const.} \tag{3.12'}$$

It is noteworthy that in this case we have succeeded in giving a solution to the problem, apart from specifying the numerical constant.

It is interesting to compare this conclusion with the results of an analytical solution, which can be obtained with no difficulty. Let us formulate the problem: we must find the field of the variable ϑ defined by the equation

$$\frac{d^2\vartheta}{dx^2} + \frac{W}{\lambda} = 0$$

with the boundary conditions $\left.\begin{array}{l}x=+l\\x=-l\end{array}\right\}\vartheta=0$ and the additional special requirement that the field be symmetrical with respect to the median plane $x=0$.

The general integral of the equation is obtained as

$$\vartheta=-\frac{W}{2\lambda}x^2+C_1x+C_2,$$

where C_1 and C_2 are undefined integral constants.

From considerations of field symmetry, we obtain immediately that $C_1=0$. On the basis of the boundary conditions, we find $C_2=\frac{Wl^2}{2\lambda}$. Thus the solution of the problem becomes

$$\vartheta=\frac{Wl^2}{2\lambda}\cdot\left[1-\left(\frac{x}{l}\right)^2\right]$$

or

$$\frac{\vartheta}{\frac{l^2W}{\lambda}}=\frac{1}{2}\left[1-\left(\frac{x}{l}\right)^2\right].$$

It is clear that this solution is a special case of Eq. (3.12). For the median plane of the plate, at $x=0$ (the plane of symmetry), we have

$$\frac{\vartheta}{\frac{l^2W}{\lambda}}=\frac{1}{2}.$$

Thus the constant in Eq. (3.12′) is equal to 1/2.

It can be said that the results of the analytical solution are in complete agreement with the conclusions obtained by using the generalized analysis.

* * *

We have considered a very wide circle of different problems in the theory of heat conduction, encountered various difficulties, and shown that the theory of similarity is a very useful instrument of investigation when used validly. In all cases up to now it has been assumed that the physical properties of the medium do not change, and hence the physical properties have been treated as constant quantities. However, this is no more than an approximation, and under certain circumstances could lead to serious errors. In these cases it is necessary to take into account the manner in which the physical constants vary with temperature, and obviously this can have an important effect on all the previous results. A whole new problem arises as to the methods to be used (within the limits of the theory of similarity) for expressing the variation of the physical properties with temperature.

Clearly this is a problem of prime importance, but it is not unique to the theory of heat conduction. We will therefore not go into it further here, since it will be more convenient to discuss it on a wider basis. We will return to this problem later when we are able to include the process of heat transfer in flowing media in the discussion.

B. The Motion of a Continuous Medium

17. PHYSICAL REPRESENTATION. SYSTEM OF FUNDAMENTAL EQUATIONS. CONDITIONS FOR UNIQUENESS OF SOLUTION

Now let us consider some of the processes which occur in moving media. Such processes are very important in a wide variety of technical fields and are also of great theoretical interest. They have been the objects of careful and systematic investigation for many decades. From this background there has arisen a circle of extremely difficult problems which have had an important effect on the formation and development of our present method of investigation.

The processes with which we will be concerned here are of various types, differing in their physical natures. However, they all occur in flowing media and, as a result, are in some way connected with motion. It is therefore valid to consider first of all the flow process itself in a medium. We will term the moving media *fluids*, including in this term both *liquids* and elastic fluids, or *gases*.

Thus we will consider the problem of fluid motion. It must be noted that here we have an objective which is not as wide as in the previous case. Earlier it was shown by a simple example (as an illustration to the general concepts given earlier) how the whole course of the investigation leads finally and with logical inevitability to the necessity of using the method of generalized variables. Therefore in considering the problem of the temperature field in a solid body we dealt in considerable detail with the initial stages of the investigation which were valid in analyzing this problem. Here, everything referring to the statement of the problem is given briefly and in outline, and we concentrate on the method of analysis.

In the case of fluid motion, the fundamental variable is the velocity. As a result, it is necessary in an analytical investigation to set up equations for determining the velocity field of the moving fluid. These equations must be sought among the general laws of mechanics. Actually it is possible to set up the *dynamic equations of fluid motion* (i.e., the equations defining the system of forces acting on the fluid), which, as mentioned before, represent a special form of d'Alembert's principle, and are known in hydrodynamics as the Navier-Stokes equations.

We must explain, of course, that not all the forces can be expressed directly in terms of the velocities of the fluid elements. One of the forces depends on the pressure distribution. Hence there are at least two important variables in the present problem: velocity and pressure.

In addition, some of the fluid properties which have an important effect on its motion depend on temperature. However, to avoid making the problem too complicated it will be assumed that the effects of fluid property changes can be neglected (this is the same as was assumed earlier in the investigation of the temperature fields in solid bodies, where the quantities specifying the properties of the medium were taken as constants, in the strictest sense).

Thus the problem consists in determining the two variables, velocity and pressure, as functions of the coordinates and time, and so the dynamic equations of motion must be supplemented by another equation. As the second equation, we use the *continuity equation;* as we mentioned earlier, this is a way of expressing the law of conservation of mass. Let us consider both equations.

Let us divide the fluid in the flow region into three-dimensional elements of such small size that within them the values of all the quantities can be regarded as being uniform (i.e., their spatial variations can be neglected). Initially let us concentrate our attention on the fact that the fluid flows continuously through an element isolated in this way, and let us set up the relationships which the effects due to this must satisfy, according to the law of conservation of mass (in all cases, relativistic effects will be neglected). This leads to the equation

$$\frac{\partial \rho}{\partial t} + \operatorname{div}(\rho \vec{w}) = 0, \qquad (3.13)$$

where t is the time;

ρ is the fluid density;

$\vec{w}$ is the velocity.

However, we have decided for the present to regard all the fluid properties, including the density, as constants. Equation (3.13) therefore assumes the form

$$\operatorname{div}\vec{w} \equiv \frac{\partial w_1}{\partial x_1} + \frac{\partial w_2}{\partial x_2} + \frac{\partial w_3}{\partial x_3} = 0. \qquad (3.13')$$

Equation (3.13) is the general form of the continuity equation, while (3.13′) is its special form corresponding to the assumption that the density is constant (in this case the fluid is termed incompressible).

Now let us consider the process of fluid motion from another point of view. By fixing an elementary volume, we isolate in a sense the element of fluid mass filling it at the given moment. Let us investigate the motion of such a fluid element and consider the corresponding dynamic model. It is clear that the motion is due to forces of two categories.

In the first category are the forces caused by external force fields acting on the fluid mass as a whole. Only the gravity force will be considered from this category of forces. Such other fields as electrostatic or magnetic, for example, will be considered in special problems; although they are currently of great interest, they are not typical of the general problem of fluid motion. On the contrary, it is always necessary to take into account the gravity force, since systems cannot be isolated from the earth's gravitational field (the state of weightlessness represents an exceptional case).

The second category includes all the forces by means of which the surrounding fluid affects our isolated element. Pressure and internal friction (viscosity) are examples of such forces.

No other forces acting on the isolated element have been discovered. Therefore, by combining these with the inertial force, according to d'Alembert's principle, we obtain a group of forces which are in equilibrium. Thus the four forces—gravity, pressure, internal friction, and inertia—must be related by an equation which expresses the fact that these forces form an equilibrium system, and this means that their vector sum must be zero. In this way we have arrived at a general scheme for the dynamic equations of motion of continuous media.

Now let us consider how the individual terms of the equation are set up, bearing in mind that all the operations refer to an elementary fluid mass, but the equation as a whole refers to a unit volume of the fluid.

The gravity force acting on the element is obviously given by the product of its mass and the acceleration of the gravity force. However, all the terms in the equation of motion refer to unit volume, so the expression for the gravity force must contain the mass per unit volume, i.e., the fluid density, as a factor. Obviously, when an incompressible fluid is in motion the gravity force is a constant quantity.

The expression for the inertial force must take the form of the product of the elemental mass and the acceleration caused by the motion of the element in the velocity field and by the changes in this field in the course of time. Hence for an element with a given constant mass, represented in the equation by the density, the inertial force depends only on the distribution of the velocity in space and on its changes with time.

The expression for the force of internal friction is obtained on the basis of Newton's law. This law permits us to determine the force of internal friction from any given distribution of the velocity in space if the appropriate characteristic property of the fluid is known; this is termed the coefficient of internal friction (or dynamic viscosity). The viscosity behaves like a constant quantity (like the density) in the equation for the force of internal friction.

Thus, assuming that the density and viscosity of the fluid are known, one of the four forces acting on the element is given as a constant quantity, and two are determined by the given velocity field. The fourth force cannot be given in analogous form: the pressure force does not depend directly on the velocity distribution, but on the pressure distribution. The equations of motion therefore contain the pressure in explicit form, which means that they contain two variables: velocity and pressure. This fact, as mentioned earlier, forces us to include the continuity equation.

Thus we have arrived at the following schematic form for the dynamic equation of motion:

$$\vec{G} + \vec{P} + \vec{F} + \vec{I} = 0, \tag{3.14}$$

where $\vec{G}$ is the gravity force;

$\vec{P}$ is the pressure force;

$\vec{F}$ is the inertial force.

$\vec{I}$ is the inertia force.

Now let us express each of these terms in the equation in terms of the primary variables. For the gravity force, we can write

$$\vec{G} = g\rho, \tag{3.15}$$

where $\vec{g}$ is the accelerating force of gravity;

ρ is the fluid density.

The pressure force is expressed very simply in terms of the pressure:

$$\vec{P} = -\operatorname{grad} p. \tag{3.16}$$

The friction force can be represented in the form:

$$\vec{F} = \mu \nabla^2 \vec{w}, \tag{3.17}$$

where μ is the coefficient of internal friction of the fluid (the dynamic viscosity);

$\nabla^2 \vec{w}$ is a symbol which must be regarded as a vector with projections $(\nabla^2 w_1,\ \nabla^2 w_2,\ \nabla^2 w_3)$, since here the operator ∇^2 operates on a vector.

Obviously the inertial force is equal to the product (with the sign reversed) of the fluid density and the acceleration of the fluid element being considered. The acceleration of the element is a complex effect, since changes in the velocity of an element during its motion may arise for two different reasons. At any given point, the velocity changes because of the unsteady-state nature of the field. In addition, changes in velocity occur as the element moves from one point in the field to another, due to nonuniformity of the field. The total change in velocity is made up from these two changes, which are usually referred to as the *local* and *convective* changes (they possess the property of simple additivity, since the changes being considered occur during an infinitely small interval of time). Hence the total acceleration is defined as the sum of two components: the local component $\frac{\partial \vec{w}}{\partial t}$ and the convective component $(\vec{w} \text{ grad}) \vec{w}$.

The special symbol $\frac{D}{dt}$ is used to denote derivatives evaluated along the trajectories of the individual elements of the moving medium, and these are termed the *individual* or *substantial derivatives**. It follows that

$$\frac{D\vec{w}}{dt} \equiv \frac{\partial \vec{w}}{\partial t} + (\vec{w} \text{ grad}) \vec{w}.$$

It should be noted that all the relationships which are derived here for velocities remain valid when other quantities which define the state (or properties) of the moving element are being considered. Thus it is always true that

$$\frac{D}{dt} = \frac{\partial}{\partial t} + \vec{w} \text{ grad} . \qquad (3.18)$$

Thus the dynamic equation of motion can be given in the form

$$\rho \vec{g} - \text{grad}\, p + \mu \nabla^2 \vec{w} - \rho \frac{D\vec{w}}{dt} = 0. \qquad (3.19)$$

---

*Translation Editor's Note: The substantial derivative $\frac{D}{dt}$ is commonly denoted by $\frac{D}{Dt}$.

Equations (3.19) and (3.13) together form a system which determines the unknown variables $\vec{w}$ and p. Thus we have a closed system.* However, the problem is still not in the required form, since the boundary conditions are not defined.

One of the characteristic features of a very wide circle of problems involving processes in flowing media, as well as of many other problems in mathematical physics, is that it is impossible to derive their boundary conditions from strictly mathematical considerations, except in the simplest cases, which are not of much interest. At the same time, it is absolutely necessary to know the boundary conditions in order to apply the methods of the theory of similarity effectively. Therefore, in all the cases when boundary conditions cannot be given (or, more precisely, when the problem of uniqueness of solution cannot be solved analytically), it is necessary to carry out an additional analysis, which usually reduces to a quite simple and clear physical argument, but which sometimes grows into an independent investigation on its own. Our specific case is a characteristic example of this type of problem, for which the question of the conditions for uniqueness of solution is very complex. We cannot give all the relevant discussion here in detail, but we will attempt to give a review of the physical ideas which underlie the very unusual system of the investigation; this will enable us to be sufficiently clear here.

Of the four forces appearing in Eq. (3.19), three ($\vec{G}$, $\vec{F}$, and $\vec{I}$) are defined quite independently of any properties of the pressure field. One of these is given as a constant, while for the two others only the velocity distribution about the point being considered is important. However, if three of the four forces are fixed, the fourth is defined in a quite single-valued manner. Thus a given

---

*Strictly speaking, there are four variables to be determined: the pressure and three velocity components. Four equations are available for this purpose, since the vector equation (3.19) can be broken down into three equations of the same type for the components of the forces applied to the element.

velocity field corresponds to only one possible distribution of the pressure force.

This discussion leads us to the conclusion that the boundary conditions need only define the velocity field in some way, but not the pressure field. Thus we can say tentatively that it is possible to give only the kinematic conditions of the process. The dynamic conditions cannot be given: they must be determined. A more detailed consideration shows that this statement can be derived with full rigor. Consequently, we can take it as established that *the boundary conditions cannot contain any information as to the pressure.* In this case, we can arrive at two very important conclusions: 1) it is impossible to give a parametric value for the pressure, and so the dimensionless pressure must be formed as a group (and not as a simple ratio); 2) the pressure cannot appear in any independent dimensionless groups, and as a result, groups containing pressure cannot serve as arguments in the generalized equations.

In addition, it is easy to see that in general the absolute pressure is unimportant in this problem, and that only differences of pressure (pressure drops) are important. Since the boundary conditions do not involve pressure, it comes into consideration through the fundamental equations. In these equations, furthermore, it appears only as a differential quantity, from which it follows that the absolute value of the pressure plays no part in the process. It is useful to remember that we applied a similar argument to the temperature in the investigation of the problem of heat redistribution in a solid body. However, the situation is not entirely the same in the two cases. In contrast to the pressure, the temperature occurs in the boundary conditions, and we had to state specially that it appeared in the form of a difference in the boundary conditions, so that its absolute value was unimportant.

The whole argument is valid only to the extent that it is possible to neglect the effect of pressure on density. If there is a significant dependence of density on pressure (i.e., if the fluid cannot be regarded as incompressible), the situation changes considerably. Under these conditions, the pressure appears in the equations not only explicitly, but also indirectly through the density. As a result of this, the assumption that the kinematic conditions of the process are independent of pressure no longer holds. In this case the boundary conditions must contain characteristic values for the pressure field as well as the velocity field. Here the absolute pressure (rather than the pressure drop) becomes important, since the density is determined by the absolute pressure.

Thus, in the investigation of processes in incompressible fluids the pressure appears in the generalized equations only as an unknown variable, and always in the form of a ratio of the pressure drop Δp to an appropriate group of variables which functions as a reference scale for the pressure. The position is quite different with respect to velocity.

In general, two forms of motion differing in origin are possible: forced motion, caused by the action of externally applied processes, and free motion (natural convection), caused by nonuniformities in the density field. Spatial density differences can arise only as a result of one of the two following causes, however: either there must be physical nonuniformities in the medium (mixtures of two fluids, of solid particles or gas bubbles in a liquid, etc.), or there must be a change in the density of the medium due to temperature or pressure. Here we will study the case of homogeneous media, and it is assumed also that they possess constant physical properties (which includes a constant density). Therefore our attention will be confined for the present to forced motion, and so all of what follows will refer to this case. The case of free motion will be considered later.

Thus we have to solve the problem of what the boundary conditions include for problems of forced motion. If the motion is forced, we will be considering in all cases either the flow of fluids in channels (the internal problem) or the flow of fluids around solid bodies (the external problem), and the action of the external agency causing the fluid to flow is characterized by the rate of passage of the fluid into the system. Different methods are used for determining the fluid velocities at the boundaries of the system. In study the internal problem, the fluid flow rate (i.e., the quantity of fluid passing through the system in unit time) is usually given; in the external problem, it is customary to give the velocity of the undisturbed stream which flows past the submerged solid body. At the moment these shades of difference are not important. What is important is that the velocity is defined in one way or another relative to the stationary elements of the system (the channel walls, surface of the body). In principle there is no difference between this problem and that of a solid moving in a stationary fluid. In this case the velocity of the solid is given relative to the fluid elements sufficiently far (infinitely far in theory) from its surface.

We see that in all cases the conditions of the problem define a certain velocity which is characteristic of the problem being investigated; this important special feature of the problem of forced fluid motion is quite general. These velocities given directly by the boundary conditions arc naturally used as the reference scales. If the parametric value of the velocity is denoted by w_0, the relative velocity will be $\frac{w}{w_0}$. The velocity must appear like this in the form of a relative variable as an unknown quantity in the generalized equation which gives the properties of the velocity field. The parametric value of the velocity which is used as a reference scale here can also appear in the similarity group which relates this quantity, which is characteristic of the process as a whole, to the other parameters.

18. FORCED FLOW. DIMENSIONLESS GROUPS. HOMOCHRONICITY GROUP AND REYNOLDS AND FROUDE NUMBERS. THE EULER NUMBER

Now let us pass on to the determination of the dimensionless groups corresponding to the fundamental equation of the problem being studied. The left-hand part of Eq. (3.13) contains a single homogeneous operator. Consequently it gives no dimensionless groups. The left-hand part of Eq. (3.19) consists of four operators, three of which, corresponding to the forces $\vec{G}$, $\vec{P}$, and $\vec{F}$, are homogeneous. The expression defining the inertial force represents a nonhomogeneous operator. Let us consider this one in more detail.

The inertial force is proportional to the substantive derivative of the velocity with respect to time. This means that the inertial force field depends on the spatial and temporal changes of velocity, which together appear as a single effect. At each moment of time at each point there is a definite relationship between the local and convective velocity changes, though the velocity is changing according to some complex law which is unknown to us. We can easily obtain a characteristic measure of this ratio for the process as a whole. This measure is obviously the power group corresponding to the substantial derivative $\dfrac{D\vec{w}}{dt}$. Thus

$$\frac{(\vec{w} \text{ grad}) \, \vec{w}}{\dfrac{\partial \vec{w}}{\partial t}} \rightarrow \frac{w_0 t_0}{l}.$$

The group obtained here establishes the relationship between the parametric value of the time (t_0) and its characteristic value $\left(\dfrac{l}{w_0}\right)$. The quantity t_0 is given directly by the conditions as the period determining the rate of development of the external influences. The characteristic period of time $\dfrac{l}{w_0}$ determines the rate of the changes which occur in the system as a result of the motion of the medium. This quantity has a very simple physical significance. It represents the interval of time during which an element moving with a constant velocity w_0 (known from the conditions) travels a distance equal to the characteristic dimension l.

This expression, which is formed by the ratio of two time periods, is a very important characteristic of the process. It is important to note that the form of this expression (in agreement with its significance) is completely independent of the parameters of whichever moving element it contains. In all cases this expression has the same form; it does not characterize the individual properties of the field of the given parameter (velocity, temperature, density, etc.), but rather it characterizes *the properties of the motion of the medium itself.*

This group must be regarded as one of the homochronicity groups from the significance of the relationship it expresses. From this point of view it is analogous to the Fourier number, so the arguments given earlier in our analysis of the part played by $\dfrac{at_0}{l^2}$ remain valid here also, apart from obvious changes (i.e., to make them applicable to the process of fluid motion rather than heat conduction). In particular, it is clear without further discussion that when the conditions of the problem do not define a characteristic period t_0, then $\dfrac{w_0 t}{l}$ assumes the sense of a dimensionless time (in which the characteristic time period $\frac{l}{w_0}$ plays the part of a reference scale) and will appear in the generalized equation as an independent variable. The problem can also be stated so that the time becomes the unknown quantity (we may have to determine some interval of time, for instance, the time for one revolution of a motor, the period of shedding eddies from the surface of a blunt body, the period for bubble separation from a heated surface during boiling of a liquid, etc.). Under these conditions the group $\frac{w_0 t}{l}$ must appear in the generalized equation as a function of the corresponding dimensionless argument. Particularly in the last case the complex $\frac{w_0 t}{l}$ is known as the Strouhal number (Sh).

The expression $\frac{w_0 t_0}{l}$ $\left(\text{or } \frac{w_0 t}{l}\right)$ is the only group which owes its origin to the unsteady-state nature of the process. Therefore, in determining the form of the groups corresponding to the dynamic equations of motion it is convenient to start from the form of these equations which refers to the case of steady-state motion. Obviously the system of groups obtained in this way is directly applicable to steady-state conditions. This is extended to the unsteady state by combining these groups with the homochronicity group.

Thus, let us consider the equation

$$g\rho - \operatorname{grad} p + \mu \nabla^2 \vec{w} - \rho (\vec{w} \operatorname{grad}) \vec{w} = 0. \tag{3.19'}$$

The left-hand part of this equation contains four homogeneous operators and therefore gives three groups. Clearly the operators can be associated in pairs in any way. However, it is easy to see how much more convenient it is to use a definite order of combination which is fixed once and for all. In practice, the following groups are firmly established in the present-day literature:

$$\pi_{IG} = \frac{w_0^2}{gl}, \quad \pi_{PI} = \frac{\Delta p}{\rho w_0^2}, \quad \pi_{IF} = \frac{\rho w_0 l}{\mu}.$$

On the basis of earlier arguments, we can conclude that only two of these three groups—π_{IG} and π_{IF}—are similarity groups, and so appear in the generalized equation as arguments (playing the part of generalized parameters). These are termed the Reynolds number:

$$\pi_{IF} \equiv \mathrm{Re} = \frac{w_0 l}{\nu}, \tag{3.20}$$

where $\nu = \frac{\mu}{\rho}$ is the kinematic viscosity, and the Froude number:

$$\pi_{IG} \equiv \mathrm{Fr} = \frac{w_0^2}{gl}. \tag{3.21}$$

The group π_{PI}, which includes the quantity Δp, represents a dimensionless form of an unknown variable (the pressure drop), in which the product ρw_0^2 serves as the reference scale; the latter has a very simple physical sense, being twice the dynamic pressure calculated with the parametric value of the velocity. This group is called the Euler number:

$$\pi_{PI} \equiv \mathrm{Eu} = \frac{\Delta p}{\rho w_0^2}. \tag{3.22}$$

Obviously the generalized equation gives the Euler number as a single-valued function of the dimensionless independent variables and the dimensionless groups.

It should be noted that the quantity ρw_0^2 is used as a reference scale for setting up the dimensionless form of any dynamic effect occurring in a stream. For example, the dimensionless form of the *friction stress** is given in the form $\frac{\tau}{\rho w_0^2}$.

---

* The friction stress is the name given to the specific (referred to unit area) tangential force which acts at any point in a stream on a plane oriented along the direction of flow; it is defined as $\tau = \mu \dfrac{\partial w}{\partial n}$, where n is the direction of the normal to the plane.

19. THE PHYSICAL SIGNIFICANCE OF THE REYNOLDS NUMBER. LAMINAR AND TURBULENT FLOW REGIMES. MOLECULAR AND EDDY TRANSFER MECHANISMS

The groups obtained from the dynamic equation of motion are measures of the ratios of the forces acting on the moving fluid and determine the important properties of the process.

The Froude number characterizes the relative value of the gravity force. It is therefore important when gravitational effects play a noticeable part in the process (motion of ships, flow over weirs). However, under certain circumstances the effects caused by the action of the gravity forces are so insignificant that they can be neglected. Thus, for instance, we can omit the action of the gravity forces if the motion occurs in a horizontal plane. In practice also the gravity forces need not be considered in the general case of forced motion of light fluids (gases). Under these conditions the Froude number degenerates and disappears from among the arguments. Sometimes this type of motion (i.e., motion in which gravity forces have no effect) is termed pure forced motion.

The Reynolds number represents the ratio of the inertial force to the force of internal friction. It can be regarded as the most important characteristic of this type of process since the properties of a fluid stream depend very fundamentally and importantly on the ratio between the inertial and internal friction forces.

In order to understand the essence of this problem clearly, it is necessary to bear in mind that the moving fluid is influenced by disturbances which penetrate into the stream from outside and hinder the development of a calm and orderly form of flow. These disturbances originate at the walls bounding the system. They are also introduced in the stream entering the system. The disturbances have to be regarded as phenomena of random origin which are completely foreign with respect to the stream and in no way related to the mechanism of the process. However, the effects arising in the fluid due to the action of the disturbances develop in a regular manner and in the final analysis are stipulated by the mechanism of the process.

The forces of internal friction have a calming action on the flow, tending to make the stream conform as closely as possible to the channel. They resist any development of disturbances which disrupt the simple form of the flow dictated by the geometry of the channel. They must therefore be regarded as a factor tending to counteract the effects of the disturbances.

The inertial forces play a diametrically opposite part. It is not difficult to show that any disturbance of an orderly flow promotes the formation of inertial forces, which in turn act on the stream in a disturbing manner, sustaining and amplifying the disordered state of the motion. Consequently we can look upon the inertial forces as a factor tending to make the effects of the disturbances worse.

Thus, disturbances present in the stream fall under the influence of two opposing effects, one of which tends to suppress them, while the other tends to amplify them. The ultimate development of the process depends on the intensities of these effects. If the friction forces predominate, the disturbances do not propagate, are localized and damped out, and clearly can have only an unimportant local effect which can have no influence on the flow as a whole. On the other hand, if the inertial forces predominate, the disturbances grow, propagate, and spread over the whole fluid, leading to important changes in the nature of the flow. We conclude, therefore, that under the same external conditions (under the action of the same external disturbances) the flow may assume one of two quite different forms. One of these is distinguished by its orderly nature and the simplicity of its properties. The motion of the fluid as a whole and that of any arbitrarily small section of it are in complete agreement. If the system is in the steady state, the conditions at every point in it will remain strictly constant. The configuration of the walls determines the direction of the flow at any point in the stream at any moment of time.

In contrast, the second form of flow is characterized by a high degree of irregularity which makes the properties very complicated. In this case the picture of the fluid flow as a whole by no means determines the nature of the motion of some small element in it. An apparently steady-state flow, which, for instance, manifests itself in the form of a constant flow rate, does not exclude the possibility of continuous changes in the conditions at any point. This is due to the fact that the quantities which determine the state of the fluid motion pulsate continuously about their mean values. The flow as a whole is oriented along the walls (if only in the sense that the fluid entering the channel ultimately flows out of it, having passed along its whole length). At the same time there can be movements in quite different directions at various points in the stream.

This diversity of properties, which at first sight are contradictory and incompatible with one another, causes great difficulty in considering the process and makes it difficult to represent physically. Recent investigations are based on the concept of a real flow as the synthetic result of two flows: a smooth, steady-state flow and a pulsating flow which is superimposed on the main flow and which imparts the characteristic ambiguousness of properties to the whole flow. The smooth flow is characterized by the fields of the mean values of all the quantities. The actual (true instantaneous) value of any quantity is regarded as the sum of this type of mean value and the pulsating component. We cannot set up a complete model for the physical nature of the pulsating flow. In any case the appearance of pulsations is related to the motion of finite fluid masses (in the form of individual fluid eddies) which move across the main flow, i.e., between regions having different values of the fluid parameters. The laws of motion of the fluid elements (the conditions for their appearance and decay, their interaction with the main flow, their natural dimensions, the lengths of the paths they flow) are still not clear in many respects, though some of the problems have been studied in great detail.

Thus there are very appreciable differences between the two possible types of flow as well as in the fundamental properties of the process. These forms of flow are distinguished from one another as the laminar (Latin lamina, a stripe) and the turbulent regimes (Latin turbulus, an eddy). The problem of which regime—laminar or turbulent—will be established under given conditions obviously depends on the relationship between the inertial force and the force of internal friction.

The special role of the ratio of the inertial force to the force of internal friction is largely explained in this way. The ratio of these quantities is denoted as Re. It is therefore possible (at least in principle) to relate the problem of the flow regime to the determination of a numerical value for this similarity group. In connection with this, it is useful to remember the general remarks which were made in discussing the concept of similarity groups as measures of the ratios of the intensities of physical effects.

In characterizing the Reynolds number as a measure of the ratio of the inertial force to the force of internal friction, we do not mean that the numerical value of Re defines this ratio as a quantity exactly proportional to it. Such a view of the role of Re as a quantitative characteristic of the process would not only be incorrect, but also without physical significance, since the ratio of the forces can have very different values at different points in the stream. We must also recognize that under certain circumstances it is no longer possible to draw quantitative conclusions on the basis of the value of Re. However, it is valid to regard the quantity Re as a quantity which correctly characterizes the relationship between the forces, for instance, in the sense that in comparing streams with significantly different values of Re, a larger value of the ratio between the inertial and internal friction forces corresponds to a larger value of this number under ordinary, natural conditions of development of the process. As a result, the argument given above on the conditions for the appearance of the different forms of flow can be given a quantitative character, related to the numerical value of Re.

This leads us to the following very simple picture of the dependence of the flow properties on the quantity Re. As the value of Re becomes smaller the friction force gradually predominates. Very small values of Re correspond to a region with a very stable laminar regime, since any disturbance occurring in the stream is rapidly localized and destroyed. As Re increases, the conditions become less favorable for this to occur. At some value of Re, usually termed the critical value, the stability of the laminar regime is lost, and the motion becomes turbulent. Larger values of Re correspond to the zone of established turbulent flow. It stands to reason that the concepts of "large" or "small" values of Re are purely relative. It is impossible to find any basis for determining the critical value of the Reynolds number $Re = Re_{cr}$ directly from the fact that this number is a measure of the ratio of the inertial force to the friction force.

The critical value Re_{cr} is determined from experiments for each specific case. Thus, for example, in the very simple case of motion along a straight circular tube, a very accurate value of 2300 has been found on the basis of numerous experiments (in the number the tube diameter is used as the characteristic dimension, and the mean velocity as the characteristic velocity). This means that in this case (flow in a straight circular tube) the laminar regime is stable at values of Re less than 2300.

From all that has been said above, it follows that the Reynolds number is a quantitative indication which makes it possible to judge which regime of flow will exist. The importance of this will be apparent if it is borne in mind that when the flow regime changes all the quantitative laws of the process change; this important rearrangement is due to the change in the physical mechanisms of the phenomena occurring in the stream. In the final analysis, these phenomena are due to the motion of fluid elements in a direction perpendicular to the wall. This motion leads to interactions of the stream with the solid bodies, since the fluid elements moving towards or away from the wall serve to transmit the properties which are important for the process.

In laminar flow all parts of the fluid move along the wall. As a result, the interactions between the stream and a solid body must consist only of phenomena of a molecular nature (travel of molecules in thermal motion, spatial vibrations).

On the other hand, in the turbulent regime, which is characterized by random motion, there is motion of the fluid (finite masses of it, not just separate molecules) in all directions, particularly in the direction perpendicular to the wall. Thus in turbulent flow the interaction between a fluid and a solid body is due to these processes of the motion, since there are now macroscopic, eddy forms of transfer, the intensities of which are not of the same order of magnitude as the molecular forms (see the note on page 151).

Thus the molecular mechanism of the interaction between the stream and body corresponds to small values of Re. During the transition at the critical value of Re the very much more powerful eddy mechanism is added, and the intensities of all the phenomena increase greatly.

20. THE GENERAL FORM OF THE DIMENSIONLESS RELATIONSHIPS FOR VARIOUS CASES OF FORCED MOTION OF INCOMPRESSIBLE FLUIDS. PURE FORCED MOTION. DEGENERATION OF THE REYNOLDS NUMBER. SELF-SIMILARITY

We have thus explained the physical significance and roles of the similarity groups which appear as arguments (parameters) in the generalized velocity and pressure equations. We can now pass on to a consideration of the equations themselves. In their most general form, they are:

$$\frac{w}{w_0} = f\left(\frac{t}{t_0}, \ \frac{x_1}{l}, \ \frac{x_2}{l}, \ \frac{x_3}{l}; \ \frac{w_0 t_0}{l}; \ \text{Re, Fr}; \ P_1, P_2, \dots\right) \quad (3.23)$$

and

$$\frac{\Delta p}{\rho w_0^2} = \varphi\left(\frac{t}{t_0}, \ \frac{x_1}{l}, \ \frac{x_2}{l}, \ \frac{x_3}{l}; \ \frac{w_0 t_0}{l}; \text{Re, Fr}; P_1, P_2 \dots\right). \quad (3.24)$$

If a parametric value of time is not defined in the problem conditions, the equations become

$$\frac{w}{w_0} = f_1\left(\frac{w_0 t}{l}, \ \frac{x_1}{l}, \ \frac{x_2}{l}, \ \frac{x_3}{l}; \text{Re, Fr}; P_1, P_2, \dots\right) \quad (3.23')$$

and

$$\frac{\Delta p}{\rho w_0^2} = \varphi_1\left(\frac{w_0 t}{l}, \ \frac{x_1}{l}, \ \frac{x_2}{l}, \ \frac{x_3}{l}; \text{Re, Fr}; P_1, P_2, \dots\right). \quad (3.24')$$

For a steady-state process, we will have

$$\frac{w}{w_0} = f_2\left(\frac{x_1}{l}, \ \frac{x_2}{l}, \ \frac{x_3}{l}; \text{Re, Fr}; P_1, P_2, \dots\right) \quad (3.25)$$

$$\frac{\Delta p}{\rho w_0^2} = \varphi_2 \left(\frac{x_1}{l}, \frac{x_2}{l}, \frac{x_3}{l}; \text{Re}, \text{Fr}; P_1, P_2, \dots \right). \qquad (3.26)$$

Finally, under conditions of steady-state, pure forced flow,

$$\frac{w}{w_0} = F \left(\frac{x_1}{l}, \frac{x_2}{l}, \frac{x_3}{l}; \text{Re}; P_1, P_2, \dots \right) \qquad (3.27)$$

$$\frac{\Delta p}{\rho w_0^2} = \Phi \left(\frac{x_1}{l}, \frac{x_2}{l}, \frac{x_3}{l}; \text{Re}; P_1, P_2, \dots \right). \qquad (3.28)$$

The case of pure forced flow deserves special attention, since a very wide circle of problems can be reduced to it; these are of great interest both theoretically and as regards their applications (though the concept of a pure forced flow is the result of a known oversimplification of the real conditions). The characteristic and important feature of Eqs. (3.27) and (3.28) in this connection is that they contain only one similarity group, Re, as an argument. This means that the generalized individual case of steady-state pure forced flow is governed by only one quantitative characteristic which relates the parametric values of the different variables in terms of a numerical value of the group Re. In other words, there is only one requirement for pure forced flow phenomena to be similar—their values of Re must be the same.

In this connection it must be noted that the group Re, which is a measure of the ratio of the inertia and friction forces, is important only when the magnitudes of the two forces are commensurable. If one of the forces becomes negligibly small compared with the other, i.e., if Re assumes very small or very large values, this leads to the degeneration of the group (which therefore disappears from among the arguments in the two equations). This means that no quantitative characteristics remain by means of which a given generalized case can be distinguished from the mass of all the possible generalized cases (in other words, there is no limiting requirement of a quantitative nature which similar phenomena must satisfy). Under these conditions the process gains a very interesting property: when the parametric values of the primary variables (dimensions, velocity, and density and viscosity of the medium) are chosen quite arbitrarily, phenomena are obtained which belong to the same generalized case as long as the value of Re remains sufficiently small (or sufficiently large). Of course, it is assumed that the dimensionless forms of the boundary conditions are the same for all these phenomena.

Suppose that we are given some phenomenon. Let us begin by arbitrarily changing the physical properties of the medium, the dimensions of the system, and the flow velocity (all the parametric groups remain unchanged, however). As long as we do not stray from the zone of very small (large) values of Re in making these changes, all the phenomena throughout the changes will retain their similarity. The primary phenomenon is the model of all the phenomena which can be obtained from it as a result of this type of transformation.* Under these conditions we say that the processes possess the property of self-similarity, or are self-similar.

It should be noted that we have already encountered self-similarity in essence in studying the temperature fields of solids. Processes involving monotonic changes in the thermal state of a solid (heating, cooling) are characterized by only one similarity group $\left(\text{Bi} \equiv \frac{al}{\lambda} \right)$. However, whenever the boundary conditions of the third type degenerate into boundary conditions of the first type ($\text{Bi} \rightarrow \infty$), the group $\frac{al}{\lambda}$ disappears from among the arguments. This means that the generalized equation obtained for the temperature field does not contain any similarity groups. As a result the

---

*The phenomena A and B are related as a standard (original) and a model when they differ only in the scales of the quantities. In consequence, any phenomenon B can serve as the model of A provided it is similar to A (i.e., provided it is included in the same generalized case. For further details, see Chapter IV, Section B).

whole array of processes which can arise under these conditions falls into one generalized case (i.e., all these processes are similar to one another). Here we have a characteristic example of self-similarity. This feature of problems with boundary conditions of the first type is partly hidden by the fact that the time appears in the generalized equation in the form of a dimensionless variable $\frac{at}{l^2}$, and so there are certain rules for comparing processes with respect to time (see page 87).

Thus the whole region of self-similarity represents a single generalized case (a group of phenomena which are similar to one another). We will use a simple example to illustrate this concept: let us consider the case of steady-state flow of an incompressible fluid in a straight circular tube (the simplest case of the internal problem).

21. STEADY-STATE FLOW OF AN INCOMPRESSIBLE FLUID IN A TUBE. FLOW STABILIZATION. THE VELOCITY PROFILE. HYDRAULIC RESISTANCE. ROLE OF THE REYNOLDS NUMBER. VELOCITY AND HYDRAULIC RESISTANCE. THE RANGE OF SELF-SIMILARITY. FULLY DEVELOPED LAMINAR FLOW

It is natural to investigate the problem of fluid flow in straight circular tubes in terms of cylindrical coordinates: we introduce the coordinates x, the distance along the tube axis, measured from the inlet section; r, the distance from the tube axis in a direction normal to the axis; φ, the angle measured in a plane perpendicular to the axis from some given direction (e.g., from the vertical). The tube radius R will be used as the characteristic dimension. Very frequently the flow can be regarded as practically axisymmetric: in this case the angle φ disappears from among the independent variables, and so only the relative coordinates $\frac{x}{R}$ and $\frac{r}{R}$

are independent variables in the generalized equation. In addition, the similarity group Re must be included among the arguments of this equation. As regards the parametric groups, we will consider that the problem is such that each of the quantities is characterized by only one value, given by the conditions, which is important for the process. Of course, we can always set up a parametric group of a geometric nature in the form of ratios $\frac{L}{R}$, where L is the length of the tube. However, this ratio is not an argument of the generalized equation, since the length of the tube per se has no effect on the velocity distribution or pressure in the stream. It needs to be taken into account only in the sense that it is always necessary to satisfy the obvious requirement that $x \leqslant L$. The limitation set up by the development of the process along the length of the tube is important in practice only in the case of short tubes. Here it will be assumed that the tube is long enough for the process to be completely developed. We assume, therefore, that the group $\frac{L}{R}$ has quite a large value, and does not need to be taken into account in any way.

Thus the equation for the velocity field assumes the form

$$\frac{w}{w_0} = F\left(\frac{x}{R}, \ \frac{r}{R}; \ \mathrm{Re}\right). \tag{3.29}$$

The nature of the velocity distribution in space and, as a result, the form of the function F depend on the boundary conditions (obviously the initial conditions are unimportant, since steady-state flow is being considered). The boundaries of the system at which the kinematic conditions of the problem are stated are the initial cross section and the tube walls. The conditions at the walls are predetermined by the nature of the process itself, since directly at the surface of a body the velocity is always equal to zero (this is always the case except for the motion of highly rarefied gases, when the length of the molecular mean free path is of the same order as the tube diameter). Consequently, one of the boundary conditions can be written as

$$w = 0 \ \text{at} \ \frac{r}{R} = 1 \ .$$

In contrast, the velocity distribution at the inlet section can be given quite arbitrarily. Let us consider the simplest case of a constant velocity at the inlet. The second boundary condition then is:

$$\frac{w}{w_0} = 1 \ \text{at} \ \frac{x}{R} = 0 \ .$$

Thus the velocity is the same at all points in the inlet cross section. As the fluid moves into the tube $\left(\text{i.e., as} \ \frac{x}{R} \ \text{increases} \right)$ the nature of the velocity distribution over the tube cross section changes (Fig. 4).

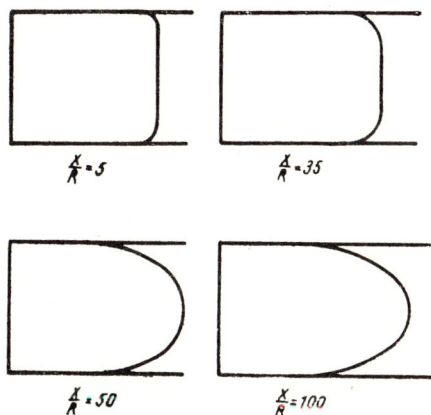

$$\frac{x}{R} = 5 \qquad \frac{x}{R} = 35$$

$$\frac{x}{R} = 50 \qquad \frac{x}{R} = 100$$

Fig. 4

The fluid elements which are in the immediate vicinity of the tube walls experience a retarding effect emanating from the surface.

This effect gradually spreads throughout the whole mass of fluid. There is an increase in the thickness of the zone near the wall in which the wall effect is felt. This zone is usually termed the *boundary layer*.

The decrease in the flow velocity in the zone near the wall must be accompanied by an increasing acceleration in the flow of the mass of fluid near the tube axis (since the flow rate of the fluid and the cross section of the tube do not change along the length of the tube). The rearrangement of the velocity field is obviously completed by the meeting of the boundary layers, after which the kinematic picture of the flow is completely developed. At the point where the boundary layers meet a velocity distribution which is characteristic for the flow being considered is established. This distribution—the *velocity profile*—remains unchanged thereafter. The manner in which the velocity field rearranges itself $\left(\text{i.e., the}\right.$ law governing the change of the boundary layer thickness $\dfrac{\delta}{R}$ as a function of the distance from the inlet section $\left.\dfrac{x}{R}\right)$, the length of the entrance region $\left(\text{i.e., the value of } \dfrac{x}{R} \text{ at which the boundary layers}\right.$ meet$\left.\right)$, and the shape of the velocity profile $\left(\text{i.e., the equation}\right.$ giving the velocity $\dfrac{w}{w_0}$ as a function of the coordinate $\left.\dfrac{r}{R}\right)$ are all important kinematic features of the process and are governed by the value of the Re number.

Let us fix on some value of Re. In this way we specify a special generalized case. Under these conditions there is only one possible velocity field which is described by Eq. (3.29) (to be absolutely correct, there is a group of mutually similar fields). Thus in kinematic respects all the properties of the process are defined by the value of the group Re with a degree of completeness and reliability corresponding to the system of fundamental equations. By giving Re various numerical values, we obtain kinematic pictures of the process development in the different cases corresponding to the numerical variants of the conditions for the problem.

Later we will return and consider the effect of the group Re on the process of boundary layer formation. Here we must confine ourselves to the following remarks. Initially the boundary layer is always laminar. It retains its laminar nature for some flow distance, but after reaching some thickness the laminar flow loses its stability. The relative value of this thickness $\frac{\delta}{R}$ and also the length at which this value is reached are determined by the value of Re.

It is not difficult to show the relationship which must exist between this length and the length of the entrance region. If $Re < Re_{cr}$, the flow in the boundary layer remains laminar over the whole length of the entrance region, and the velocity profile in the region of fully developed flow is formed as the result of laminar boundary layers merging. The flow retains its laminar nature for the whole flow distance. But if $Re > Re_{cr}$, a turbulent boundary layer will begin at some cross section within the entrance region. Under these conditions the boundary layers when they meet have already undergone a transition to the turbulent state, and the fully developed flow is turbulent.

The dimensionless pressure is given as a function of the same group of arguments. However, we must remember that the pressure has the same value over any given tube cross section. In this case the coordinate $\frac{r}{R}$ must be omitted from the arguments. The equation becomes

$$\frac{\Delta p}{\rho w_0^2} = \mathrm{Eu} = \Phi\left(\frac{x}{R}; \ \mathrm{Re}\right). \qquad (3.30)$$

Thus the problem reduces to the determination of the pressure drop between the initial and some subsequent cross section corresponding to the given value of $\frac{x}{R}$. The quantity Δp represents the sum of two terms describing effects of quite different physical natures. The first represents the pressure change caused (according to Bernoulli's law) by the change in the kinetic energy of the moving mass of fluid. When the kinetic energy is increased, the pressure falls, and when it is decreased, the pressure rises. This is therefore a reversible effect. This component is consequently called the reversible part of the pressure drop. In direct contrast to this, the second component, which gives the pressure drop due to energy dissipation (i.e., due to the transformation of mechanical energy into thermal energy), is essentially irreversible. It represents the irreversible part of the pressure drop. Under the conditions of the present problem (flow of an incompressible fluid along a channel of constant cross section) a reversible transformation of energy can occur only within the entrance region. As long as the formation of the velocity profile is still going on, with the mean velocity over the cross section remaining constant ($\overline{w} = w_0$, where the bar over the quantity denotes averaging), the kinetic energy of the fluid increases, and this has a corresponding effect on the pressure distribution along the tube. However, in the region of fully developed flow, where the velocity field is completely established, the reversible part of the pressure drop falls to zero, and the quantity Δp consists entirely of its irreversible part.

The pressure drop caused by the dissipation of energy characterizes the hydraulic resistance of the tube. Thus in the region of fully developed flow the pressure drop is a measure of the hydraulic resistance. A special quantity—*the coefficient of hydraulic (hydrodynamic) resistance** —is introduced as the characteristic of the hydraulic quality of the tube; this is very simply related to the Euler number (i.e., to the dimensionless pressure drop).

In fully developed flow the process conditions remain unchanged in all respects from one section to another (there is only a fall in the absolute pressure and, generally speaking, this has no effect on the process). It follows from this that beyond the entrance region the dissipation intensity remains constant along the tube. The longitudinal pressure gradient (an essentially negative quantity) is therefore constant, so the pressure falls linearly. Under these conditions, it is convenient to introduce a specific pressure drop. The Euler number is obviously transformed accordingly by multiplying it by the reciprocal of the relative length. This modified Euler number represents the hydraulic resistance coefficient. It must be remembered that the reference scale for the pressure drop is the dynamic pressure $\frac{\rho w_0^2}{2}$ (and not the product ρw_0^2), so that the relative length is introduced in the form $\frac{L}{d}$, where $d = 2R$ is the tube diameter.

Thus, denoting the resistance coefficient by ζ, we have

$$\zeta = \frac{\Delta p}{\dfrac{\rho w_0^2}{2}\,\dfrac{L}{d}}, \tag{3.31}$$

or

$$\zeta = 2\mathrm{Eu}\,\frac{d}{L}, \tag{3.31$'$}$$

where Δp denotes the pressure drop over the whole length of the region of fully developed flow.

*Or friction factor—Translator.

It is immediately clear that the resistance coefficient is a function only of the Reynolds number:

$$\zeta = \zeta(\text{Re}). \tag{3.32}$$

Solving Eq. (3.31) for Δp, we find

$$\Delta p = \zeta \, \frac{\rho w_0^2}{2} \, \frac{L}{d}. \tag{3.33}$$

Thus the pressure drop becomes a quantity which is proportional to the square of the velocity. However, the actual relationship between Δp and w_0 is still not established, since the proportionality factor ζ itself depends on the velocity. Obviously, the less ζ depends on Re the more closely the change of Δp as a function of w_0 approximates to a squared relationship. Let us look into this problem more closely.

We know that small values of Re, up to $\text{Re} = \text{Re}_{cr}$, correspond to the region of stable laminar flow, i.e., to the flow region in which the internal frictional force has a dominant effect. This leads to self-similarity of the process so that all its properties are independent of the numerical value of Re. Consequently, it is to be expected that all the quantitative characteristics of the flow (the law for the growth of the boundary layer along the length of the tube, the velocity distribution over the cross section) remain unchanged in relative form over the whole range of values of Re up to Re_{cr}. Both experiments and theoretical considerations completely confirm this general nature of the properties of all laminar flows. In particular, the velocity profile over the cross section is always parabolic for fully developed flow in the laminar regime. This result of a simple analytical solution has been confirmed by careful experiments.

Similar generality of the properties of laminar flow is observed in the investigation of the hydraulic resistance. It is found, in particular, that under the conditions of the laminar regime the dimensionless pressure cannot be given in the form of an Euler number. Actually the group $\frac{\Delta p}{\rho w_0^2} \equiv \pi_{PI}$ represents a measure of the ratio of the pressure force (a quantity of the same order as the force of internal friction) to the inertial force (which is negligibly small compared with the friction force). As a result, it has an infinitely large value for laminar flow, and obviously loses its significance as a generalized variable which is of importance for the process. Thus in this region the Euler number Eu, as well as the Reynolds number Re, degenerates. However, by combining the groups π_{PI} and π_{IF} in the form of the product $\pi_{PI}\pi_{IF} = \pi_{PF}$, we obtain

$$\pi_{PF} = \frac{\Delta p R}{\mu w_0} , \tag{3.34}$$

which obviously is a measure of the ratio of two forces of the same order of magnitude, and therefore serves as a suitable form for representing the dimensionless pressure in the present flow regime. It is often called the Lagrange number:

$$\frac{\Delta p R}{\mu w_0} \equiv \text{La.} \tag{3.34'}$$

Since the Lagrange number is a function of the Reynolds number in the general case, it is obvious that La has the same value for all laminar flows (since these are all processes falling in the same range of self-similarity).

Thus, for "small" values of Re, we can say that

$$\frac{\Delta p R}{\mu w_0} - \text{const.} \tag{3.35}$$

From this it is clear that there is a "first-power law" in the range of laminar flow—the pressure drop is proportional to the first power of the velocity. It is interesting to compare this result with Eq. (3.33), which, as we noted, is formally constructed from a "squared law." In essence there is no contradiction. The apparent disagreement of the two equations only indicates that under the present conditions the resistance coefficient ζ depends markedly on Re. It is not difficult to find the nature of this dependence. The condition La = const is equivalent to the requirement EuRe = const or ζRe = const.

Thus, for the whole region of laminar flow, the resistance coefficient (or, equally, the Euler number Eu) is defined as a quantity inversely proportional to Re (an analytical solution gives $\zeta = \dfrac{64}{Re}$, where the diameter d is used as the characteristic dimension). The main point here is that under the present conditions the Euler number (or coefficient ζ) is not a quantity which is characteristic of the flow, and it cannot be used as a generalized variable. It is only in conjunction with Re (which by itself is also not applicable under the present conditions) that the group La is formed as their product; this group is a generalized variable which is completely characteristic of the process. Within the laminar flow regime this group is completely independent of Re, in agreement with the general theoretical conclusions.

When the value of the Re number is increased so much that it exceeds the critical value Re_{cr}, the laminar regime loses its stability. A new regime begins in which both of the competing forces are quantities of the same order of magnitude, and consequently neither of the tendencies—towards the decay or towards the growth of external disturbances—predominates. We usually call this the *transition regime*. The process becomes extremely sensitive to external effects; this leads to a very complicated and unstable picture in which it is extremely difficult to distinguish any real lasting pattern. In general it is scarcely possible to characterize the flow properties in the transition regime by definite and sufficiently general quantitative laws.

By its very nature, the transition regime cannot have a clear upper limit. It is not correct, therefore, to attach a definite value of the Re number to its termination. Rather, we must assume that as Re increases, the nature of the flow varies continuously.

As the value of Re increases, so does the ratio of the inertial force to the force of internal friction, and there begins to be an increasing tendency for the amplification and spread of disturbances arising in the flow. At Re values of the order of 10^4 this process has gone so far that the flow can be regarded as completely turbulent. This means that at values of Re $= 10^4$ or greater it is necessary to use the quantitative laws which characterize the new properties of the flow satisfactorily (and which therefore differ considerably from the laws which are valid in the laminar flow regime).

The manner in which the properties of the flow evolve as Re is increased further can easily be predicted by the same methods used in the discussion above. Obviously there must be a gradual and continuous decrease in the part played by the force of internal friction in shaping the flow, with a corresponding decrease in the importance of the group Re. Ultimately this leads to the degeneration of Re when the inertial forces become completely predominant. Obviously this is an asymptotic type of process. It is quite impossible to think of the change in the flow properties in such a way that Re serves as a generalized argument up to some definite value (and is therefore a similarity group in the strict sense of the word) and then ceases to play this part. The actual evolution of the flow properties is such that the generalized characteristics gradually depend less and less on Re. The exact value of Re at which its effect can be neglected (in other words, the point where the region of self-similarity begins) is a matter of definition, depending on the required degree of accuracy. However, it would be wrong (as in the case of discussing the problem of where the transition region ends) to abandon the problem of determining the value of Re at which the self-similarity region begins. It is easy to follow these features of the relationships describing the properties of turbulent flows in actual relationships.

Equation (3.32) for the hydraulic resistance coefficient can be conveniently approximated by the simple power relationship

$$\zeta = A\mathrm{Re}^{-n}, \tag{3.36}$$

where A and n are constants (positive numbers) which vary with Re.

If we wish to retain this form of the relationship for laminar flow also, we have to put $n = 1$ and $A = 64$ for the whole laminar region (i.e., both the exponent and the constant factor in the equation must be regarded as constant numbers which do not depend on Re, in exact agreement with our assumption of the properties of the laminar regime as a regime of self-similar flow). In contrast, in the turbulent regime the constants in the equation have to be regarded as variables (functions of Re), for which definite values can be specified only over limited ranges of Re. Thus, at a value of Re of the order of 10^4 we have to put $n = 0.25$ (1/4). However, the "quarter-power law" is by no means universal for the whole range of turbulent flow. Its applicability is limited to the Re range from 10^4 to 10^5. In the Re range from 10^5 to 10^6 better results are obtained by putting $n = 0.21$. Over the ranges $10^6 - 2 \times 10^6$ and $2 \times 10^6 - 5 \times 10^6$ we have to use $n = 0.19$ and $n = 0.18$ respectively.

Thus, in complete agreement with our assumption that the degeneration of the Re number is an asymptotic process, we see that there is a continuous decrease in its importance, but that this decrease occurs at an even slower rate.

These features of the relationships characterizing the properties of turbulent flows are also of definite importance in the study of the velocity distribution over the cross section in the region beyond the entrance length. For fully developed flow the velocity profile equation is

$$\frac{w}{w_0} = f\left(\frac{r}{R}; \ \mathrm{Re}\right),$$

which can be successfully approximated by the power relationship

$$\frac{w}{w_{\max}} = \left(\frac{r}{R}\right)^m, \qquad (3.37)$$

where $w_{\max}$ is the velocity at the tube axis.

The exponent m is a function of Re, which is the way the effect of this group on the velocity distribution is expressed in this form of the power approximation.* For the same ranges of Re mentioned above ($10^4 - 10^5$, $10^5 - 10^6$, $10^6 - 2 \times 10^6$, $2 \times 10^6 - 5 \times 10^6$), m takes the following values: 0.143 (1/7), 0.118, 0.102, and 0.100 (1/10). It is easy to see that the rate of change of m is decreasing gradually.

We see, therefore, that the effect of the group Re on the quantities ζ (or Eu) and m (and hence on n) decreases continuously. Consequently we can always specify some value of Re above which its effect can be neglected to the required degree of accuracy, and above this value we can regard the coefficient ζ (the Euler number Eu) and the exponent m as constants. This means that beginning with this value of Re a definite and unchanging velocity distribution over the cross section and hydraulic resistance is established.

---

*A more detailed examination shows that the exponents m and n are related by the equation $(2 - n) m - n = 0$. Therefore the way in which one of them changes predetermines the behavior of the other.

The importance of this result is that this value of Re marks the beginning of the region of flow which forms a single generalized case, i.e., of a region of self-similarity. It is clear that, in agreement with what was said above, it is impossible to derive a definite lower limit for this region by strictly theoretical means, since essentially the choice of the limiting value of Re depends on the degree of accuracy which is required. The region of self-similarity has no upper limit.

A characteristic feature of this region of flow is that the "squared law" of resistance applies throughout it, so that $\zeta = $ const. Another feature is the bluntness of the velocity profile (i.e., it tends towards a rectangular configuration), corresponding to very small values of the exponent m; this behavior is typical of this flow zone. Quantitatively the bluntness of the profile, which is caused by a high intensity of eddy transport, makes itself felt in the fact that the condition $w_{max} = \overline{w} = w_0$* holds to a high degree of accuracy.

We have therefore arrived at a clear picture of the way in which the flow properties depend on the value of Re (Fig. 5). Small values of Re correspond to laminar flow, which forms the first region of self-similarity. The value Re $= 2300$ marks the upper limit of this region. Throughout the whole of this region the condition EuRe $=$ const (or ζ Re $=$ const) is satisfied, so that the hydraulic resistance is given by a first-power law.

It is impossible to set up any definite quantitative relationships for the transition region which then ensues. An Re value of the order of 10^4 can be regarded as the beginning of the turbulent flow region, for which we must use a power law for the hydraulic

---

*In the limit this equality is satisfied exactly. It follows from the equation connecting the exponents m and n that as n tends to zero, so does m. At $m = 0$ the velocity profile assumes a rectangular configuration.

resistance coefficient ζ in which the exponent is a steadily decreasing function of Re. At large enough values of Re, this region passes into the second region of self-similarity whose characteristic condition is Eu = const (ζ = const), so that the squared law of resistance is valid.

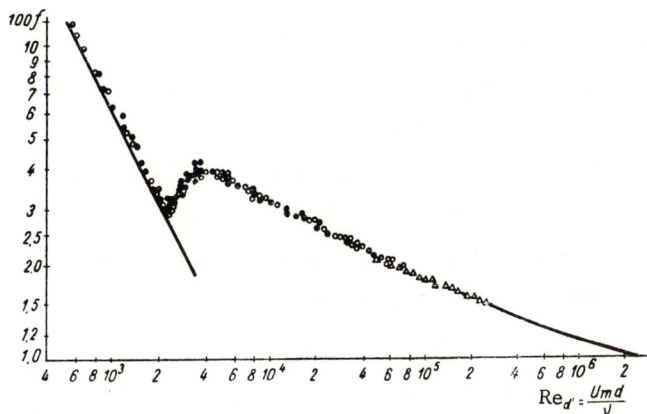

Fig. 5

The allocation of the whole of the flow between the individual regions is somewhat arbitrary. The boundaries between the regions are so unclear that we should speak of a continuous transition from one region to another rather than of a sudden change of regions. The critical value Re_{cr} is an exception to this, since it can be fixed fairly definitely. However, the transition across the critical value Re_{cr} is not necessarily related to a change in the form of flow. As we have seen, the value Re = Re_{cr} is characteristic in the sense that the laminar form of flow loses its stability there. This means that the turbulent flow regime becomes physically possible. However, the actual realization of this possibility depends on external causes rather than on the internal properties of the flow. There is only a change in the stream properties when we pass the critical value Re_{cr} if disturbances pass into the stream from outside. With respect to Re_{cr}, therefore, we can only say that at $Re < Re_{cr}$ the laminar form of flow retains its stability. Even so, certain restrictions must be placed on this statement, since at values of Re approaching Re_{cr} the laminar flow loses its pure form, characterized by the strict linearity of the trajectories of the fluid particles, which occurs at very small values of Re. The trajectories become wavy, and this effect (which can be observed visually) gradually becomes more marked as Re_{cr} is approached.

In connection with our discussion on the effect of the Reynolds number, on the one hand, and the part played by external disturbances in making the stream turbulent, on the other, it will be of interest to consider the following problem. We see that the disturbances passing into the stream from without are the primary cause of the appearance of turbulence. We have supposed that the natural properties of the stream are only important in setting up the conditions which ensure the decay of growth of these disturbances. This is based on the assumption that turbulence cannot arise from the action of the internal mechanism of the process. Let us prove that this assumption is valid. Under laminar flow conditions, the velocity components normal to the tube axis tend to zero. The expression for the inertial force therefore becomes

$$\vec{I} = \rho w \frac{\partial w}{\partial x},$$

where the subscripts on w and x can be omitted, since this does not lead to any confusion.

However, the velocity remains constant along any trajectory, which means that $\frac{\partial w}{\partial x} = 0$. In this case $I = 0$. Thus the inertial force is characteristically absent in laminar flow. At the same time, only the inertial force can lead to internal disturbances of the flow, so that the laminar form of flow can never be changed by purely internal factors.

Thus the penetration of external disturbances into the stream is a necessary prerequisite for the formation of turbulence in the stream. This means that a stream which is isolated from the action of external disturbances can never in principle become turbulent. This is an unexpected result. If we artificially set up conditions under which there is no possibility for external disturbances to reach the stream, the laminar form of flow will persist up to values of Re as high as we please. This conclusion has been confirmed experimentally. Laminar flow has been obtained at values of Re up to 10 times larger than the critical value. Of course, such flows are extremely unstable, and any accidental disturbance inevitably leads to instantaneous and vigorous turbulence (i.e., transition occurs). In a sense, this type of flow is analogous to the metastable state of matter (supercooled or superheated liquids, supercooled vapors), which can only exist due to elimination of factors which are necessary for transition of the material to the state which is more stable under the prevailing conditions (e.g., centers of crystallization, vapor formation, or condensation must be avoided).

The fact that unstable laminar flow can exist is of interest in connection with the correct understanding of the parts played by the similarity groups as measures of the ratios of physical effects and, in particular, with the understanding of the Reynolds number as a measure of the ratio of the inertial force to the force of internal friction. With specially constructed equipment, it is possible to have flows in which the effect of the inertial force is completely absent (and in which, therefore, the ratio of the inertial force to the force of internal friction has a value of zero) even at values of Re at which the inertial force would greatly predominate

under normal conditions of flow. It is obvious that in these unstable, artificially sustained laminar flows the Reynolds number
cannot be regarded as a characteristic of any part of the process,
since one of the effects which leads to the stream properties
characterized by Re is eliminated completely. This discussion is
a very useful illustration of the concept mentioned earlier in
general form that predictions of the properties of a process based
on numerical values of similarity groups become more reliable
the more exactly it is possible to specify the tendencies which are
inherent in the process and which cause its mechanism.

22. FREE FLOW. SPECIAL FEATURES IN SETTING UP THE PROBLEM AND TRANSFORMING THE GROUPS. THE GALILEO AND ARCHIMEDES NUMBERS. THERMAL CONVECTION. THE GRASHOF NUMBER

Up to now we have studied only forced fluid flows, and we have
been able to show that when applied to this problem the method of
generalized analysis is a very effective means of investigation.
Now let us consider how we can apply these methods to the study
of free fluid flows. It is very instructive to compare the results.

Under forced flow conditions, we are always given some value
of the velocity from the statement of the problem, regardless of
the exact details of the situation. This feature of forced flows,
which has already been noted, is very important to us, since it
specifies the roles of the numbers Re and Fr as similarity groups.
These quantities can be introduced only as arguments when we are
studying processes involving forced motion. They may, of course,
degenerate, in which case they disappear from the solution.
However, they can never be converted into dimensionless forms of
unknown variables.

In contrast, in the investigation of free flows it is impossible
(due to the nature of the process itself) to give any value of the

velocity in advance. In this case the moving fluid does not pass through the system and does not intersect the boundaries. A free flow is confined by the limits of the system, and the properties are specified by the physical conditions arising within it. Everything that can be said directly about the velocity (and that must be reflected in the problem conditions) can be reduced to the single statement that the velocity must be equal to zero at the surface of a stationary solid body. In addition, if the sources giving rise to the motion are localized, the velocity can be regarded as practically zero at a large enough distance from the sources.

As we will see, problems of free motion are always set up so that the conditions do not define any value of the velocity which might be used as a parameter. The consequences of this are clear. The dimensionless velocity cannot be given in the form of a simple ratio, since we have no reference scale. The Froude and Reynolds numbers become dimensionless variables of the dependent type. The velocity field equation contains the groups $\dfrac{w}{\sqrt{gl}}$ or $\dfrac{\frac{w}{\nu}}{l}$ as the unknown variable. The two groups $\mathrm{Re} = \dfrac{wl}{\nu}$ and $\mathrm{Fr} = \dfrac{w^2}{gl}$ disappear from among the arguments (generalized parameters). However, by dividing the two expressions containing the velocity, we can obtain a combination in which the velocity does not appear. Thus:

$$\frac{\mathrm{Re}^2}{\mathrm{Fr}} = \frac{gl^3}{\nu^2}. \tag{3.38}$$

This group contains only defined quantities. It is therefore a similarity group and appears in the generalized equation as an argument. It is usually called the Galileo number:

$$\frac{gl^3}{\nu^2} \equiv \mathrm{Ga}. \tag{3.38'}$$

The significance of the group Ga as a measure of a ratio of forces which operate in free motion processes is not immediately clear. Let us rewrite Eq. (3.38) in the following way:

$$\mathrm{Re}^2 : \mathrm{Fr} \equiv \pi_{IF}^2 : \pi_{IG} = \pi_{IF}\pi_{GF} . \qquad (3.38'')$$

It is now obvious that the group Ga can be represented as a product of two groups, one of which is a measure of the ratio of the inertial force to the force of internal friction, while the other is a measure of the ratio of the gravity force to the force of internal friction.

The Galileo number is not the only one which characterizes the process of free flow. It is easy to see that under these conditions groups of the parametric type must also be introduced. If we consider the case of free flow resulting from the differences in the physical properties of two materials which compose the system (a discussion of this very simple case will illustrate the main features of the problem), these groups will be ratios of the physical constants which are of importance for the process.

The group corresponding to a property denoted by u is written as $\dfrac{u'}{u}$, where the prime quantity refers to the second material. It is not difficult to establish which constants are important for the process. Obviously the primary quantity is the density, since nonuniformities in the density field are the reason for the appearance of free convection. Hence the group $\dfrac{\rho'}{\rho}$ must be included among the arguments under all circumstances when we are considering either mixtures of two fluids, or fluid media including other phases in a dispersed state (solid particles, or bubbles of gas or vapor). In addition, in the first case (two fluids) the course of the process will depend on the ratio between the fluid viscosities, so that under these conditions it is necessary to introduce the group $\frac{\mu'}{\mu}$ $\left(\text{or } \frac{\nu'}{\nu}\right)$.

The quantity $\frac{\rho'}{\rho}$ can easily be reduced to a form in which it has the significance of a quantity which is very characteristic of the process of free motion. Actually it is clear that it can be replaced by the equivalent expression $\frac{\Delta\rho}{\rho}$, where $\Delta\rho = |\rho - \rho'|$. However, $\frac{\Delta\rho}{\rho}$ is the relative buoyancy, i.e., the ratio of the buoyant force $\vec{g}\Delta\rho$, which arises due to the difference in density between the two media (and which is the resultant of the gravity force and the hydrostatic pressure), and the gravity force $\vec{g}\rho$, acting on a "unit" of fluid. In this form, it represents a parametric group characterizing the nonuniformity of the density field. In this case, we have to deal with the set of groups $Ga = \frac{gl}{\nu^2}$ and $\frac{\Delta\rho}{\rho}$, to which must be added the group $\frac{\mu'}{\mu}$ when appropriate. Obviously this combination can be replaced by another equivalent to it: $\frac{gl^3}{\nu}$, $\frac{\Delta\rho}{\rho}$, and $\frac{\Delta\rho}{\rho}$. This last group is often termed the Archimedes number:

$$\frac{gl^3\Delta\rho}{\rho\nu^2} \equiv Ar. \tag{3.39}$$

It is easy to see that the Archimedes number is a special modification of the Galileo number, and can be obtained directly if initially we introduce the buoyancy force $\vec{g}\Delta\rho$ into the equation instead of the gravity force $\vec{g}\rho$. In this case we obtain $\frac{\rho w^2}{gl\Delta\rho}$ instead of the Froude number, and so Eq. (3.38″) becomes

$$Re^2 : \frac{\rho w^2}{gl\Delta\rho} = \pi_{IF}^2 : \pi_{IAG} = \pi_{IF}\pi_{AGF}. \tag{3.39'}$$

The physical significance of this result is clear without further comment.

In conclusion, let us say a few words about the process of *thermal convection.*—This is an extremely common form of free motion, and is of interest theoretically as well as having very important applications. The special feature of this case is that the nonuniformity of the density field is caused not by differences in the properties of dissimilar materials, but by differences in the state of a single material at different points in space. Essentially the effect of thermal expansion of materials is the basis of the thermal convection phenomenon (in consequence, in considering this process we have to give up the assumption that the properties of the medium are constant, at least as regards the density). The density field is determined by the temperature field, which in turn is formed by the action of the heat transfer process.

Thus under free flow conditions the two problems—that of fluid motion and that of heat transfer—cannot be separated, but form different facets of the same overall problem. Naturally it is impossible to study this problem in any detail without taking the heat transfer phenomena into account. For this reason our discussion of thermal convection here will be limited to some simple considerations, and we will return later to this problem for a more detailed study.

Let us consider the simplest case of a single source of disturbance giving rise to motion in a uniformly heated medium of infinite extent. This elementary arrangement, corresponding to the pure form of the process uncomplicated by any additional effects, is well suited due to its simplicity for analyzing the essential features of the process.

A medium with a temperature T_0 has placed in it a body with some other temperature T_1, which remains unchanged throughout the process (this additional condition is necessary, since we wish to consider steady-state motion). The space filled by the fluid is very large compared with the dimensions of the body. As a result of the interaction between the body and the medium, velocity and temperature distributions are set up (and hence a density distribution also). Here we will not go into the nature and mechanism of this interaction, since this will be the subject of detailed discussion in the investigation of heat transfer under conditions of free motion.

The equations giving the manner in which the variables change in space may be quite complicated, and are determined by the specific physical conditions of the process. However, with regard to the temperature, it is clear that its value must fall between T_0 and T_1. With this in mind, we reason as follows. First we introduce the coefficient of thermal expansion β, defined by the equation

$$\beta = \frac{1}{V} \frac{dV}{d\vartheta}.$$

Here $V = \dfrac{1}{\rho}$ is the specific volume of the medium; and $\vartheta = |T - T_0|$, where T is the varying value of the temperature.* In this case, the buoyant force (referred to unit volume) at a given point in the field can be given in the form $g\rho\beta\vartheta$ to a sufficiently good approximation when the value of ϑ is small. This determines the form in which the similarity group containing the buoyant force must be written. Naturally a value of the temperature difference given by the conditions of the problem must be introduced into the expression for the group:

$$\vartheta_1 \equiv |T_1 - T_0| \equiv \Delta T.$$

---

*More precisely, $\beta = \dfrac{1}{V} \left(\dfrac{\partial V}{\partial \vartheta} \right)_p$, where the subscript p denotes that the derivative is taken at constant pressure. This condition is almost always satisfied under free flow conditions, since the pressure changes are very small compared with the absolute pressure.

Thus the Archimedes number assumes the form $\frac{gl^3\beta\Delta T}{\nu^2}$. This "thermal modification" of the Archimedes number is known as the Grashof number:

$$\text{Gr} \equiv \frac{gl^3\beta\Delta T}{\nu^2}. \qquad (3.40)$$

Obviously the parametric group $\frac{\Delta\rho}{\rho}$ is replaced by the expression $\beta\Delta T$.

We have shown that under conditions of thermal convection the generalized velocity equation contains the Grashof number and $\beta\Delta T$ as arguments. However, this does not mean that the investigation has been completed. The coefficient of thermal expansion β is a function of temperature, and so it is not clear which value of it should be used in the group. Some uncertainty arises in considering any of the physical constants which depend on the temperature (ν, for instance).

The methods to be used in investigating processes occurring in media with varying physical properties present very important problems, and we will consider them further. So far we have assumed that the temperature change has been quite small, and have been satisfied with the degree of accuracy obtained by assuming that the physical properties remain constant. Even this way of looking at the problem does not make it possible to neglect changes in the density (as the primary cause of the process of free motion), since thermal convection occurs even with quite small density differences.

The results obtained so far are not sufficient, because up to now we have not taken into account in any way the physical circumstances of the process which relate to the heat transfer phenomena. The temperature field (and consequently the whole hydrodynamic picture) depends significantly on these phenomena. It is obviously necessary to add similarity groups characterizing the process of heat transfer. We will now pass on to the study of this process.

We must note in advance that all the cases of free motion considered here are processes of gravitational convection, since they are a result of the action of the earth's gravitational field. These processes are physically impossible in a weightless state. However, this does not preclude the possibility of other free motions occurring under the action of other types of force fields. A typical example, which is of great technical importance, is thermal convection in a centrifugal force field. This field can be several times stronger than the earth's gravitational field. For instance, the centrifugal acceleration occurring in modern turbines may reach a value of 10^4 g or more. Thermal convection may clearly reach extremely high intensities under these conditions. This effect has been successfully applied to the cooling of gas turbine blades by fluids circulating in closed channels.

C. Heat Transfer in Moving Media

23. THE MECHANISM OF HEAT TRANSFER IN MOVING FLUIDS. SYSTEM OF FUNDAMENTAL EQUATIONS

The physical effect which manifests itself as heat transfer between a solid and a surrounding medium is essentially a process of heat transfer in a direction normal to the surface of the body. In convective heat transfer, i.e., in heat transfer between a solid and a fluid, this process is carried out by a transfer of heat in a direction perpendicular to the surface of the solid (for the sake of brevity, this will be termed a *transverse* migration).

This transverse migration of heat transfer always exists, and, as a result, heat transfer between the solid and the fluid occurs under all conditions. However, there are various forms in which this migration can occur. The properties of the heat transfer fluid may also differ.

We have already discussed in general form the problem of the interaction between a body and the fluid surrounding it (page 121). To summarize the results of this discussion, we arrive at the conclusion that we must distinguish between two forms of heat transfer: molecular and eddy transfer. It is therefore necessary to relate a heat transfer process to a definite mechanism of heat transfer. A correct statement of the problem must include the operative mechanism of heat transfer for the specific conditions.

The transfer of heat in the form of a molecular process occurs under all conditions regardless of the state of motion of the medium. These processes occur by various means in fluids, but for the case of solids they are combined under the term of heat conduction. The processes are at their simplest in gaseous media. In this case the agents transporting the heat are the individual molecules (or atoms) of the gas. These take part in a random thermal motion, i.e., in a motion which occurs in all possible directions, and hence they travel also in the direction normal to the surface of the submerged body. The mechanism of this process is simple and its main features are quite clear; it has been investigated quantitatively and thoroughly.

The physical nature of the thermal conduction process in liquids is much more complicated. At present many aspects of this process are still not clear, since we do not possess a sufficiently complete theory of the liquid state. According to recent concepts, it is probable that the properties of both gases and solids occur simultaneously in some form in a liquid. It is therefore clear that the molecules (or atoms) alone are not the only agents for transferring heat in a liquid. It is probable that quantities of acoustic energy also perform a similar function; these arise as a result of the thermal oscillations of a molecular lattice which forms and breaks up periodically. Other more complex forms of transfer may also occur.

In the present state of the problem, it has so far been impossible to set up general and rigorous quantitative definitions. However, in all cases heat conduction in gases and liquids is a low-intensity process except in metallic liquids, in which thermal conduction occurs in the form of electron conduction, so that the transfer agents are free electrons. This form of heat transfer occurs naturally in stationary media and under the conditions of the laminar fluid flow regime. It will be remembered that a characteristic feature of laminar flow is that the direction of the fluid motion as a whole coincides with the direction in which any separate part of the fluid moves. Therefore there is generally no macroscopic motion in the transverse direction under laminar flow conditions, and the only possible form of heat transfer must occur by processes of a molecular nature.

In direct contrast, the eddy mechanism begins to play a part in turbulent streams; under conditions of fully developed turbulence this mechanism is enormously more effective than the molecular one. Thus, heat transfer in a turbulently flowing fluid is the result of a combination of the effects due to phenomena of different physical natures. The relative importance of each of the two forms of transfer depends on the state of the fluid motion. The molecular mechanism is completely dominant in streams with fully developed turbulence.* However, turbulence dies away as the surface of a solid body is approached, and very close to the solid surface the effect of the molecular mechanism begins to become quite noticeable. In the immediate vicinity of the surface the

*It is necessary to exclude some media—liquid metals—which are characterized by high thermal conductivities (and hence, by very small values of the Prandtl number Pr). In these media molecular heat transfer plays an important part under all conditions.

turbulence is damped out to such an extent that molecular processes predominate. This part of the stream is usually termed the *laminar sublayer*. The simple laws of thermal conduction are applicable in this zone as long as there are no effects on the surface of the body which disturb the fluid flow.

We will base our consideration of heat transfer between a fluid stream and a solid body on this representation of the physical mechanism of the transfer process. Now let us deal with the problem of the fundamental equations for the case of heat transfer in a flowing medium.

The fundamental equation of the theory of heat propagation in a fluid stream is expressed by the law of conservation of energy, modified for the conditions under which an element of the moving fluid interacts with the mass of fluid surrounding it. Considering the case of a steady-state process, this equation can be given as follows (all the terms refer to unit volume):

$$\vec{w} \operatorname{grad}\left(T + \frac{w^2}{2c_p}\right) = a\nabla^2\left(T + \frac{\nu}{a}\frac{w^2}{2c_p}\right), \qquad (3.41)$$

where, in addition to the symbol already used, we have introduced the symbol c_p for the heat capacity at constant pressure (this is the same heat capacity that appears in the expression for the thermal diffusivity $a \equiv \frac{\lambda}{c_p\rho}$).

This equation applies rigorously only in the case of an ideal gas, i.e., a material which satisfies the equation of state

$$pv = RT \qquad (3.42)$$

or

$$p = \rho RT, \qquad (3.42')$$

where $v = \frac{1}{\rho}$ is the specific volume of the material and R is the gas constant. It can also be applied to other media with sufficient accuracy, however.

Equation (3.41) contains two unknown variables, T and $\vec{w}$. It must therefore be considered in conjunction with Eqs. (3.13) and (3.19), which determine the quantity $\vec{w}$. Together these equations form a system of three equations with three unknowns ($\vec{w}$, T, and p), so that the system is closed. However, this is true only to the extent that all of the properties of the medium denoted in the equations by the physical properties ρ, λ, μ, and c_p can be regarded as constant. This reservation has been made throughout. With respect to the density, it is only in the case of liquids that this reservation does not contradict the statement of the problem (we have seen that even here precautions are needed in studying free motions). In the case of gases the density must be regarded as a function of the temperature and pressure, and changes in density can only be neglected if the temperature and pressure remain within a quite narrow range.

The form of this function is given by Eq. (3.42); under normal conditions (temperatures which are not too low and pressures which are not too high) this equation can also be applied to real gases with sufficient accuracy for practical purposes. Thus we have arrived at a system of four equations [(3.13), (3.19), (3.41), and (3.42)] with four unknowns (T, $\vec{w}$, p, ρ). It must be noted that under certain conditions the physical constants change so rapidly that it becomes quite impossible to regard them as constants (for instance, c_p changes very rapidly near the critical point). This greatly complicates the problem.

A characteristic feature of Eq. (3.41) is that it contains the velocity in the form $\dfrac{w}{2c_p}$, which is summed directly with the temperature in the left-hand part of the equation, and occurs in combination with the factor $\dfrac{\nu}{a}$ in the right-hand part. In the case of liquids the quantity $\frac{w^2}{2c_p}$ is almost always (there are only a few exceptions) negligibly small compared with T, even after being multiplied by the ratio $\frac{v}{a}$, which has a value much larger than unity for most liquids. The second terms within the parentheses can therefore be neglected for the case of liquid motion. In the case of gaseous media, high-velocity flows for which the quantity $\frac{w^2}{2c_p}$ is of the same order of magnitude as T are quite possible and are frequently used in modern industry. However, in many fields flows of moderate velocities are more characteristic.

The study of heat transfer was started and developed initially in the period when the main practical interest was in heat engines and equipment limited to low-velocity processes. It is only much more recently that the theory of heat transfer has been revised and freed from its initial limitations under the pressure of recent technical requirements. Following the historical development of the theory, we will assume that the second terms in both parts of the equation can be neglected (in the case of gases the ratio $\frac{v}{a}$ is a fraction differing only little from unity). With these assumptions, Eq. (3.41) becomes

$$\vec{w}\,\text{grad}\ T = a\nabla^2 T. \qquad (3.43)$$

24. THE DIMENSIONLESS GROUPS FOR THE PROBLEM OF HEAT TRANSFER IN MOVING FLUIDS. RELATIONSHIPS BETWEEN THE PECLET AND REYNOLDS NUMBERS. THE PRANDTL NUMBER AND ITS PHYSICAL SIGNIFICANCE. GENERAL FORM OF THE TEMPERATURE FIELD EQUATION UNDER FORCED FLOW CONDITIONS

Equation (3.43) contains two homogeneous operators. There is therefore one corresponding group which can be easily derived by the usual methods:

$$\frac{\vec{w}\operatorname{grad} T}{a\nabla^2 T} \longrightarrow \frac{w_0 l}{a}.$$

All the quantities appearing in this group are known from the conditions under forced flow conditions. It is therefore a similarity group. It is named the Péclet number:

$$\frac{w_0 l}{a} \equiv \mathrm{Pe}. \tag{3.44}$$

It is not difficult to indicate the physical significance of the Péclet number. If the reduction operation is rewritten as

$$\frac{\rho c_p \vec{w}\operatorname{grad} T}{\lambda \nabla^2 T} \longrightarrow \frac{w_0 l}{a},$$

we can see that the group formed in this way represents a measure of the ratio of the intensity of heat transfer by convection to the intensity of heat transfer by thermal conduction. In other words, the group Pe is a measure of the relative roles of the convective and molecular mechanisms of heat transfer.

On comparing Pe with the groups obtained earlier, its similarity to the Reynolds number $\frac{w_0 l}{\nu}$ is immediately apparent. The numerators of the two groups are identical. The denominators represent physical constants: in the first case the denominator is the thermal diffusivity, and in the second the kinematic viscosity. It is not difficult to explain these general properties of the two groups in a formal manner. It is sufficient to remember that the Reynolds number is obtained as the final result of comparing the operators $(\vec{w}\operatorname{grad})\vec{w}$ and $\nu\nabla^2\vec{w}$, which differ from the operators in Eq. (3.43) in two respects only: 1) they apply to the variable $\vec{w}$ rather than to T (this has no effect on the structure of the group), and 2) the second contains the factor ν rather than a (this is an important difference, and appears in the groups).

However, it is much more important to note the common properties of the two groups, as these throw light on the similarities of their structures. It is not immediately clear how physical effects related to one another can be expressed in terms of quantities which are so different in their physical meanings as the Reynolds number (which is a measure of a force ratio) and the Péclet number (which characterizes the relative intensities of convective and molecular heat transfer). The important point is that forces do not appear directly in the dynamic equations of motion. In agreement with the model of the process, they are represented by momentum fluxes. Essentially the Reynolds number is a quantitative measure of the relative values of the momentum fluxes due to convective and molecular mechanisms. Thus the Reynolds and Péclet numbers represent exactly analogous characteristics of the processes for the transfer of heat (Pe) and for the transfer of momentum (Re). There is therefore a fundamental physical reason for this similarity in the structures of the groups.

In this sequence of ideas it is noteworthy that Re and Pe differ only with respect to physical constants characterizing molecular transfer: on replacing ν by a the Reynolds number is transformed into the Péclet number. The groups have no structural differences due to the features of the convective transfer. It is useful to isolate the quantities which lead to the differences between these groups as a separate group. This is easily done by taking the ratio:

$$\frac{Pe}{Re} = \frac{\nu}{a}.$$

The ratio $\dfrac{\nu}{a}$, which we have encountered already in Eq. (3.41), is termed the Prandtl number

$$\frac{\nu}{a} \equiv Pr. \tag{3.45}$$

The special feature of the Prandtl number is that it is formed from only single physical constants and therefore as a whole has the property of a dimensionless physical constant. It therefore has a definite value for each medium, assuming that the physical properties do not vary. As mentioned earlier, the Prandtl number for gases is not much different from unity. On the other hand, for liquids it is greater than unity, and for higher viscous liquids it may reach very large values (10^3 or more). Liquid metals are exceptions; they have extremely small values of Pr (of the order of 10^{-2} or even 10^{-3}).

The Prandtl number is of interest in other respects also. It follows from its structure that the constants v and a are equivalent in the special sense of this term which is characteristic of the theory of similarity (see page 43). This is the first time that we have encountered a case in which a given quantity is equivalent not to a group of quantities, but to another single but physically different quantity. This can be compared with the fact that the quantities v and a play exactly similar parts, each in its own application.

The thermal diffusivity, as we have shown, characterizes the ability of the medium to react to the passage of a heat flux by changing its temperature. In the same way, the kinematic viscosity characterizes its ability to react to the passage of a momentum flux by changing its velocity.

In order to conclude our discussion of the problems related to the origin of the Péclet number and its relationship to the other groups, we should note that it can be obtained in a quite different way.

If we do not confine ourselves to the case of a steady-state process the homochronicity groups must also be introduced, since there is a specific relationship characterizing the rate of development for each of the processes (fluid flows and heat distributions in the fluid). We have already determined these groups. The first is written in the form $\frac{w_0 t_0}{l}$, and the second in the form $\frac{a t_0}{l^2}$. It is easy to form a combination of these two groups which does not include the time. This combination also clearly represents a group, but it differs from the initial groups by the fact that it remains valid under steady-state conditions also. Combining these groups in the form of the ratio of the first to the second, we obtain $\frac{w_0 l}{a}$, which is none other than the Péclet number.

Thus when we are investigating a phenomenon which occurs in a moving fluid and includes a heat distribution process, the dimensionless groups must include the quantity $Pe \equiv \frac{w_0 l}{a}$ (which can be replaced by the dimensionless parameter $Pr \equiv \frac{v}{a}$). Under forced flow conditions Pe is a similarity group, as are the other groups which contain the velocity (Re, Fr).

The Prandtl number can obviously be regarded as a similarity group under all circumstances. In the case of steady-state pure forced flow, therefore, we obtain two groups, namely Re and Pe, or, more usually, their equivalent Re and Pr. For instance, the generalized equation for the temperature distribution in a fluid interacting with a solid under conditions of pure forced flow is written in the form

$$\frac{\vartheta}{\vartheta_0} = f\left(\frac{x_1}{l}, \frac{x_2}{l}, \frac{x_3}{l}; Re, Pr; P_1, P_2, \dots\right), \qquad (3.46)$$

where ϑ_0 is the difference between two values of the temperature T_1 and T_0 given by the conditions, and ϑ is the difference between the value of the temperature at the given point and T_0.

In the very simple but quite typical case of the flow of a fluid of given initial temperature along a straight circular tube of constant radius R with a constant wall temperature, we have

$$\frac{\vartheta}{\vartheta_0} = F\left(\frac{x}{R}, \frac{r}{R}; \text{ Re, Pr}\right). \tag{3.46'}$$

Here the wall temperature is regarded as the zero (T_0), and ϑ_0 is defined as the difference between the initial fluid temperature and the wall temperature.

In the region of self-similar flow Re disappears. It need hardly be mentioned that under these conditions it is impossible to replace Pe by Pr.

25. THE TEMPERATURE FIELD EQUATION UNDER FREE FLOW CONDITIONS. DEGENERATION AND AMALGAMATION OF GROUPS. TURBULENCE DURING THERMAL CONVECTION

Thermal convection is a characteristic case of free motion; under the conditions of thermal convection the group Pe ceases to be an independent group and, like Re, reverts to a dimensionless form of an unknown variable (w). A new group must therefore be introduced in the form of Pr. The generalized equation therefore contains two groups: Gr and Pr (the quantity $\beta\Delta T \equiv \beta\vartheta_0$ must obviously be regarded as a group of the parametric type). We find, therefore, for the velocity distribution:

$$\frac{wl}{\nu} = F_1\left(\frac{x_1}{l}, \frac{x_2}{l}, \frac{x_3}{l}; \text{ Gr, Pr, } \beta\vartheta_0; P_1, P_2, ...\right), \tag{3.47}$$

and for the temperature distribution:

$$\frac{\vartheta}{\vartheta_0} = F_2\left(\frac{x_1}{l}, \frac{x_2}{l}, \frac{x_3}{l}; \text{ Gr, Pr, } \beta\vartheta_0; P_1, P_2, ...\right). \tag{3.48}$$

Clearly the dimensionless velocity in the left-hand part of Eq. (3.47) could equally well be given in the form $\frac{wl}{a}$. We must remember that the expression $\beta\vartheta_0$ appears twice in the right-hand part of the equation, once in Gr and once as an independent group. However, experiments show that in many cases the effect of this quantity is so small that it can be neglected.

Now let us consider a problem which is characteristic of the theory of similarity as a method of quantitative investigation. Suppose that the inertia force is so small compared with the other forces that it can be neglected. In this case it is apparently impossible to form Re and Fr. This means, however, that it becomes impossible to formulate Ga either (and as a result, Ar also, and its modification Gr) or Pr. Therefore both similarity groups of Eqs. (3.47) and (3.48) disappear.

However, it does not follow at all from this that the equations can contain no groups of this type under these conditions. Actually, when the terms corresponding to the inertial force are omitted from the equation of motion it is still possible to form π_{GF}, which includes the velocity. Obviously there must be some combination of this and the number Pe which gives a group which does not contain the velocity. It is easy to find this combination. It can be seen from Eq. (3.35″) that multiplying π_{GF} by π_{IF} ($\equiv$ Re) converts it into Ga. However, in this case, by combining π_{GF} as a product with Pe = PrRe, we again obtain a similarity group (equal to GaPr). It follows ultimately from this that the disappearance of the inertial force leads to the amalgamation of Gr and Pr into a single group in the form of the product GrPr. The problem naturally arises now as to what changes are necessary in the solution in the opposite case, when the friction force is negligibly small. Under these conditions Re disappears, and therefore so do Ga and Pr, which are derived from it. However, in contrast to the case above, the Froude number Fr is retained. By analogy with (3.38) we therefore find directly:

$$Pe^2 : Fr = Pr^2 Ga.$$

It follows that the final equation must contain one similarity group of the type $GrPr^2$ instead of the two independent groups Gr and Pr.

The results obtained above are very instructive. The concept of the degeneration of similarity groups (as an indication of the unlimited attenuation of certain physical effects of importance in the process) is characteristic of the general theory; here it has been given a clear and concrete significance. The case being considered here is particularly interesting since the group appearing as an argument in the final equation is a product of the very groups which have just degenerated. In complete agreement with the theory, however, the disappearance of one effect leads to the amalgamation of groups; the procedure for forming the similarity groups is quite clear.

Hence for free flow there is a definite relationship between the relative value of the inertial force and the structure of the arguments in the final equation. Small relative values of the inertial force correspond to the type GrPr, and large values to the type $GrPr^2$. It is natural to assume that the law characterizing the intermediate region (in which the inertial force and the force of internal friction are of the same order of magnitude) can be approximated by a relationship based on an argument of the type $GrPr^m$, where $1 < m < 2$. All these conclusions can be checked experimentally.

Let us start with the following simple discussion. It is clear from Eq. (3.39′) that the effect of the inertial force must increase as the value of Gr increases. As a result, a group of the form GrPr applies in the zone of small values of this argument. (It must be noted that the combination of Gr and Pr in the form of the product cannot obscure the overall picture, since Pr is practically a constant for any given fluid, and so on passing from one fluid to another there is merely a known change in the numerical value of this parameter.) Experiments have confirmed this conclusion beyond a shadow of doubt. The effects of Gr and Pr can be represented over a wide range by the product GrPr. At very large values of this product, however, better results are obtained by using an argument of the type $GrPr^m$. The form $GrPr^2$ corresponds to very high-intensity processes, which are encountered quite infrequently in practice.

The following fact is noteworthy in connection with this discussion. The possibility of combining Gr and Pr in the form of a simple product GrPr is a sufficient basis for concluding that we can neglect the effect of the inertial force on the development of the process. There are no problems in estimating the dynamic conditions of the process in this way. However, great caution must be used in passing on from this to a determination of the kinematic picture. To some degree the usual representation of the nature of the relationship between the relative values of the forces acting in the fluid and the flow regime loses force here, though this representation is quite valid under conditions of pure forced flow. The situation is gravely complicated by the fact that here there are sources of disturbances within the fluid itself which may lead to perturbation of the flow. These sources are the buoyant forces.

Reduced to the bare physical essentials, the process of free motion of a medium consists in a redistribution of the elements of the medium which have different densities, in such a way as to reach a more stable state. The state increases in stability as the potential energy of the system in the gravity force field decreases (or, in a wider sense, as the potential energy of the system in the external mass force field decreases).

A solid placed in a medium at another temperature is always a source of disturbances to stability (except in special cases, such as a horizontal surface with the warm side turned down or with the cold side turned up). The primary cause of the free flow process' occurring is the thermal interaction between the body and the medium. The elements of the medium adjacent to the surface of the body assume its temperature. The process of heat propagation into the fluid by conduction begins, and a temperature field arises. The temperature changes lead to changes of density, and these in turn disturb the initial density distribution corresponding to the hydrostatic pressure field. It is characteristic of the new density distribution that in some areas elements of greater density will occur at positions higher than elements of lower density. Such a distribution is unstable, since it does not correspond to the least possible value of the potential energy in the system. Under these conditions equilibrium can be disturbed by an insignificant inter-action; the buoyant force effect appears, and a process of density redistribution occurs: the warmer elements rise, and the cooler ones move downwards. In practice all sorts of random external effects act continuously, and the process assumes the form of a continuous mutual displacement of fluid elements.

Thus the redistribution of the fluid elements is directed towards the reduction of the potential energy of the whole collection of elements. There is therefore a continuous evolution of energy which is used up in sustaining the phenomena connected with its dissipation. It is easy to see that this leads to conditions which favor the transition of the flow into the turbulent form. Under these conditions it would obviously be foolish to try to solve the problem of the flow regime on the basis of elementary arguments as to the relative values of the inertial force and the force of internal friction. It is all the more impossible to regard the whole region in which an argument of the form GrPr is applicable as a region of laminar flow. Actually, an optical study of the flow patterns under free flow conditions has shown that even in the very simple case of flow along a vertical wall (when the conditions are most favorable for the retention of the laminar regime) turbulence appears long before it ceases to be valid to represent the group in the form of the product GrPr.

26. HEAT TRANSFER BETWEEN A MOVING FLUID AND A SOLID. DIMENSIONLESS FORM OF THE HEAT TRANSFER COEFFICIENT. THE NUSSELT NUMBER. GENERAL FORM OF THE EQUATION FOR THE INTENSITY OF HEAT TRANSFER

Equations (3.46) and (3.48) give us a great deal of very general information about the thermal conditions of the process. If the functions in these equations are known (by means of numerical solutions or experiments) we will have obtained as much information as it is possible to obtain from the statement of the problem. It is easy to see what a huge amount of work is required to carry the solution to this degree of completeness. However, in fact such complete information is often superfluous. In most cases there is no real need to carry out a detailed investigation of the temperature field in the stream since the practical requirement is the determination of a heat transfer coefficient. In fact, a more important and more frequently encountered technical problem consists in finding the quantity of heat transferred between a solid body and a fluid surrounding it for a temperature driving force which is given or can be determined relatively easily. The solution of this problem presents no difficulties if the heat transfer coefficient is known. On the other hand, it is often necessary to investigate the temperature distribution in constructional elements which operate as heat exchangers (checking the construction for

the limiting attainable temperatures; determination of thermal
stresses in constructional elements). Thus, although the temperature field of the fluid is not of interest from a practical point of
view, the temperature field of the solid body interacting with it is
of great importance. As we have seen (pages 58-59), a prerequisite for solving the temperature field in a solid body is
almost always that the value of the heat transfer coefficient must
be given directly.

Thus in one way or another technically important problems are
connected with the problem of the heat transfer coefficient. Of
course, it is always possible to obtain the heat transfer coefficient
from a complete solution of the temperature distribution problem
in the fluid. However, as we have already explained, the important
point is to be able to find the heat transfer coefficient without
investigating the temperature field of the fluid as a whole. It is not
difficult to set up a plan for solving this problem by using appropriate experiments. By definition, the heat transfer coefficient is
given by the equation

$$\alpha = \frac{q}{\Delta T} \, .$$

The experiments must therefore allow us to determine separately
each of these quantities—the heat flux density q and the temperature driving force ΔT. Experimental methods for the separate
determination of these two quantities under various real conditions
have reached a high degree of development. The experimental
results can be represented in the form of a generalized relationship giving the heat transfer coefficient as a single-valued function
of the complete groups corresponding to it. It is easy to see that
these groups are the same as the groups in Eqs. (3.46) or (3.48)
respectively. Both quantities—the heat flux density and the temperature driving force—are completely determined by the temperature

field of the fluid and depend only on the quantities which are arguments for the temperature itself. As a result their ratio, given in relative form, must be a single-valued function of the arguments appearing in the generalized equation for the temperature.

Thus the right-hand part of the generalized equation for the heat transfer coefficient is known right away, since it is the same as the right-hand part of the temperature equation. However, there are certain difficulties in connection with representing the heat transfer coefficient itself in a dimensionless form; these are due to the fact that it is impossible to give any value of this quantity directly, so that there is no reference scale for it. Here we have a situation which we have already encountered in the general theory. We know that this situation, as such, does not lead to any difficulty, since the problem can be resolved by replacing the deficient parameter by a group of quantities equivalent to it.

The present case differs by the characteristic feature that the heat transfer coefficient does not appear in any of the fundamental equations, so we do not have available the groups from which the equivalent group of quantities should be selected. This leads to a very real difficulty. It is obviously necessary to derive at least one equation which relates to the heat transfer process in the fluid and contains the heat transfer coefficient. Such an equation can be set up on the basis of the following argument.

In the immediate vicinity of the surface of the solid body only the molecular mechanism of heat transfer is operative (page 151). The heat transfer intensity is therefore given directly by the law of pure heat conduction. Consequently, we have for the density of the heat flux crossing the surface

$$q = -\lambda \left(\frac{\partial T}{\partial n} \right)_{n=0},$$

where λ is the thermal conductivity of the fluid and n is the normal to the surface.

On the other hand, the quantity q can be represented in the form

$$q = \alpha \Delta T,$$

where q is taken to mean the absolute value of the heat flux density, since the temperature driving force is defined in terms of the absolute values of the temperature of the surface and the medium. Thus we arrive at the following equation, which outwardly is in complete agreement with Eq. (3.3):

$$\alpha \Delta T = \lambda | \text{ grad } T |_0. \tag{3.49}$$

There is only one group, $\dfrac{\alpha l}{\lambda}$, corresponding to Eq. (3.49). Structurally this group is identical to the Biot number, which of course is natural in view of the agreement between the forms of Eqs. (3.3) and (3.49). However, the present group differs from the Biot number in many respects. The Biot number contains the thermal conductivity of a solid body which is interacting with a fluid. The Biot number establishes a definite relationship between the intensity of heat transfer and the temperature distribution in the solid—it is only in this way that it can serve as a measure of the ratio between the temperature drop in the solid body and the temperature driving force.

In contrast to this, the new group contains the thermal conductivity of the fluid. It is obviously quite unrelated to the physical properties of the solid body, and does not in any way determine the temperature field which occurs in the solid. In addition to these conspicuous differences, there is another difference of great importance which is not immediately obvious.

The Bi number contains the heat transfer coefficient as a *given* quantity, which is known from the statement of the problem.

Equation (3.3) serves as a boundary condition for defining the temperature gradient at the surface of the body for a given heat transfer coefficient and temperature driving force. In complete contrast, in the statement of the problem which is now of interest to us, the heat transfer coefficient must characteristically be regarded as an *unknown* quantity.

This group $\frac{\alpha l}{\lambda}$ must be introduced into the generalized equation which is used for determining the value of α. It therefore represents a dimensionless form of the heat transfer coefficient, and is by no means a similarity group. We see, therefore, that it would be quite incorrect to identify it with the Biot number. In agreement with the terminology given earlier, the dimensionless form of the heat transfer coefficient will be called the Nusselt number:

$$\frac{\alpha l}{\lambda} \equiv \mathrm{Nu}. \tag{3.50}$$

We have therefore established the form of the generalized equation for the heat transfer coefficient. This equation defines the Nusselt number as a function of the groups which we have already found. For the most general assumptions, we will have for pure forced flow:

$$\mathrm{Nu} = F_1\left(\frac{x_1}{l},\ \frac{x_2}{l},\ \frac{x_3}{l};\ \mathrm{Re},\ \mathrm{Pr};\ P_1,\ P_2,\ ...\right), \tag{3.51}$$

for free convection:

$$\mathrm{Nu} = F_2\left(\frac{x_1}{l},\ \frac{x_2}{l},\ \frac{x_3}{l};\ \mathrm{Gr},\ \mathrm{Pr};\ \beta\vartheta_0;\ P_1,\ P_2,...\right). \tag{3.52}$$

If an unsteady-state process is being considered and the heat transfer coefficient varies with time, the relative time $\frac{t}{t_0}$ must be included among the groups as an independent variable, as well as the Fourier number $\mathrm{Fo} \equiv \frac{at_0}{l^2}$ as a parameter $\Big($in the case of forced motion, we must introduce the group $\frac{w_0 t_0}{l}$ instead$\Big)$. If no scale value is given for the time, the dimensionless form of the time becomes $\frac{at}{l^2}$ $\Big($or $\frac{w_0 t}{l}$ for forced motion only$\Big)$, and the corresponding generalized parameters disappear.

The equations obtained above are very complicated. The multiplicity of arguments makes it extremely difficult to determine the form of the functions. However, in practice it is usually not necessary to use the full, complicated equations. Quite frequently it is possible to confine our attention to the determination of the mean value of the heat transfer coefficient for the system as a whole. In this case the problem is greatly simplified, since the coordinates disappear from among the arguments. The equations are reduced to

$$\mathrm{Nu} = \Phi_1 (\mathrm{Re},\ \mathrm{Pr}; P_1, P_2, ...) \tag{3.51'}$$

and

$$\mathrm{Nu} = \Phi_2 (\mathrm{Gr},\ \mathrm{Pr},\ \beta\vartheta_0;\ P_1, P_2, ...). \tag{3.52'}$$

Equation (3.52') can be simplified still further.

As we noted earlier, the quantity $\beta\vartheta_0$ can be omitted without any noticeable loss of accuracy. In addition, it is convenient to combine Gr and Pr in the form $\mathrm{Gr}\mathrm{Pr}^m$; we can assume $m = 1$ over a wide range. Thus the equation can be written in the form

$$\mathrm{Nu} = \Psi (\mathrm{Gr}\ \mathrm{Pr}^m, P_1, P_2, ...)$$

or, for a limited but quite wide range of values of GrPr, even in the form

$$\mathrm{Nu} = \Psi (\mathrm{Gr}\mathrm{Pr}; P_1, P_2, ...).$$

27. EFFECT OF VARIABLE PHYSICAL PROPERTIES. STATEMENT OF THE PROBLEM. THE TEMPERATURE FACTOR. THE METHOD OF CHARACTERISTIC TEMPERATURES. THE EFFECT OF THE HEAT FLUX DIRECTION

Now let us discuss the effects which arise when the physical properties vary with temperature. We have mentioned the importance of this problem several times already, but its discussion has been postponed until we have collected the necessary factual material. We are now sufficiently prepared to be able to give the whole discussion a concrete aspect. Let us concentrate on processes which involve the interaction of moving media with solid bodies; the problems arising in studying these are the most difficult and physically interesting.

In setting up the problem in completely general form we must assume that all the physical constants (in addition to the density) are functions of temperature only. In this way our object is to investigate the features of the interaction between the fluid stream and the solid due to the changes in the physical properties of the medium brought about by temperature changes. These features appear in the form of various physical effects. On the one hand, there are direct effects of the changes in the properties of the medium (thermal conductivity, viscosity) on which the intensities of the processes of interest to us depend, and, on the other hand, there are secondary effects (changes in the velocity profile of a forced flow because of viscosity changes, superposition of a free motion on a forced flow) which have important effects on the course of the process.

However, regardless of the various forms assumed by the phenomena which we are considering, their primary cause is the variation of the physical properties of the medium with temperature. We must therefore discover and study those features of the processes which are the consequences of changes in the properties of the medium caused by changes in its temperature.

The theory of similarity in the general form in which we have considered it up to now has not been adapted to the solution of such problems. In developing the fundamental aspects of the theory of similarity we have proceeded from the assumption that the physical properties are constant quantities. We have taken the requirement that fields of the physical properties should be similar to mean that, in the case of a complicated system, the physical properties in its component parts should be connected by definite relationships (by expressions in the form of parametric groups). Within each homogeneous region the physical properties are assumed to be constant.

The situation is radically different in setting up the problem of the effects of temperature on the physical properties. Any physically homogeneous region in which a temperature distribution is set up is characterized by corresponding fields of the physical properties.

How does the fact that the physical properties are not constant parameters but variables which depend on the state of the medium affect the process of generalized analysis? How is the variability expressed in terms of generalized relationships?

If the quantities characterizing the physical properties of a system depend on temperature, the equations expressing the forms of their temperature dependences must be included among the fundamental equations of the process. The similarity groups corresponding to these equations must be regarded as arguments in the generalized relationships. These obviously include all the new groups connected with the variability of the physical properties. However, this answer is not sufficient. Omitting mention for the moment of the great increase in complexity of the generalized equations, which become practically useless as a result, the following must be borne in mind.

In most cases it is impossible to derive the equations giving the temperature dependences of the physical properties from general theoretical grounds. These are given by purely empirical relation-ships. These equations can differ even in their structures from one material to another (particularly for materials of widely different physical properties). Under these conditions, combining such relationships with the fundamental equations leads to the formation of different systems of equations in each case. Obviously a generalized analysis is quite impossible under these conditions, since such an analysis is entirely based on the premise that we are considering phenomena belonging to the same class (i.e., phenomena which are defined by a given system of equations). It is therefore pointless to seek a solution of this problem in a rigorous general form. We must be satisfied with approximate solutions, which are only of limited value.

One of these solutions is based on representing the physical properties in the form of a power function of the absolute tempera-ture. The power function clearly possesses the following very interesting property: a given distribution of the dimensionless group corresponds to the given distribution of the dimensionless function, or, put another way, similarity in the distributions of the arguments leads necessarily to similarity in the distributions of the functions (see also Chapter IV, Section 30). This means that with the present assumption (that all the functions are of a power type) the only similarity groups for the physical constants are those which define the absolute temperature field, and hence the new variable becomes the absolute temperature rather than the temperature difference.

The effects of the variability of the physical properties leads to this special feature, which has important consequences. Heat transfer problems are usually formulated so that the conditions are determined by two values of the temperature (e.g., the tube wall temperature and the temperature of the fluid at the inlet section). Up to now only the temperature difference has been important, so that this statement of the problem conditions has defined one temperature difference (the temperature driving force).

Now, when the absolute temperatures are important, we must assume that the problem conditions define two values of the variable. In this case a new parametric similarity group must be introduced in the form of the ratio of these values. It is easy to see the part played by this group in determining the physical conditions of the process. Under these conditions a new effect (previously excluded from consideration) becomes important, namely, the change of the physical properties with temperature. A necessary condition for distributions of the properties to be similar is that the values of the relative temperatures for the various phenomena must be equal. In other words, in a generalized individual case the ratio of the absolute temperatures given by the conditions is a new quantitative characteristic; corresponding to this (together with all the other characteristics) there are definite distributions of the relative values of the constants in space. This ratio must obviously be included among the arguments of all the generalized equations. Thus this leads to the addition of one new argument representing a temperature group of the parametric type.

Thus one possible way of extending the applicability of the generalized equations obtained earlier to the case of media with variable physical properties is to add another new argument. This argument is given by the ratio of the two values of the absolute temperature given in the conditions (the temperature of the medium and the wall temperature). This ratio is often called the *temperature factor*.

All the physical constants appearing in the groups must be referred to one of these temperatures. In principle it makes no difference which is chosen as the reference temperature. However, analysis of experimental data for gas flows of moderate velocities leads to the conclusion that it is most suitable to refer the properties to the higher temperature, since this reduces the effect of the temperature factor (or even makes it practically disappear).

This interesting effect has still not been given a rational explanation. In any case, apart from some of the details, the "temperature factor" method is based on a definite system of theoretical arguments. However, it must be remembered that this method of generalization is closely bound up with the applicability of the power formula for the physical constants, and so it rests on convincing grounds only to the extent that these formulas can be regarded as satisfactory.

For gases all the physical properties can be closely approximated by power functions over a wide range of temperatures. The generalized equations including the temperature factor among the arguments can therefore be used validly in studying gas flows.

The conditions are less favorable in the case of liquid flows, because the viscosities of liquids are powerfully temperature dependent in a complicated way. Under these conditions the fundamental role of the temperature factor as an argument in the generalized equations ceases to be valid. To overcome this difficulty, we introduce an argument into the generalized equation which is formed from the ratio of the dynamic viscosities at the temperature of the medium and at the wall temperature. It is not immediately clear that this ratio is really an argument, and it is not possible to prove this. This must therefore be regarded as an empirical form of generalization, depending mostly on chance (its use gives good results over quite a wide range of "moderate" temperatures frequently encountered in practice); its choice depends more on a superficial analogy than on fundamental physical arguments. Conditions are particularly complicated in the region immediately around the critical point. There is characteristically an extremely sharp change in the heat capacity in this region (i.e., in one of the constants which is only very slightly temperature dependent under ordinary conditions). This unexpected effect complicates the process so much that it is still not clear in just what form the generalized equations should be applied here.

From all that has been said, it follows that attempts to derive a general method for expressing the effects of variable physical properties by means of special arguments introduced into the final equations have led to rigorous results based on theory only in the case of gases.

However, this is not the only possible approach to the problem. Another form of solution is very widely used; this is based on the concept that the effect of variability of the physical constants appearing in the dimensionless complexes can be taken into account by referring the constants to some "mean" temperature which is characteristic of the process. The essential features of the method of characteristic (identifying) temperatures are explained by the following discussion.

The space occupied by the process is characterized by some temperature distribution in which the temperatures vary over the range given by the conditions. Fields of the physical properties corresponding to this distribution are established. Without an analytical solution of the problem it is impossible to forecast how this complicated change of a whole group of quantities will affect the final equations. However, if all of a series of temperatures (included within the range given by the conditions of the problems) affect the development of the process in one way or another, then for each distribution there must be some mean temperature (i.e., a temperature lying within the given temperature range) which has the characteristic that its substitution into the final equation gives a result which adequately takes into account the effect of the whole multiplicity of temperatures. If we can give a general method for determining this temperature (i.e., if we can find how to express it in terms of the given temperatures), we shall have completely solved the problem of how to take temperature changes into account.

In principle this is a completely valid approach to the problem. However, it cannot be realized systematically in practice. A reliable basis for the correct choice of a method for averaging the temperature could be obtained only from a complete analytical solution. Therefore the various proposals for methods of determining the mean temperature must be based directly on the general representation of the physical conditions of the process.

If we bear in mind the extremely complicated internal relationships between the physical properties and the final expressions in terms of which the generalized solution of the problem is given, it becomes clear that this type of argument is inevitably superficial to some extent, and cannot provide a firm enough foundation for setting up a reliable formula. It is no exaggeration to say that these arguments merely suggest a method for determining the mean temperature rather than place this method on a firm footing. Of course, this does not mean that it is useless to attempt to select a mean temperature. In many cases it is possible to set up convenient working formulas in this way, and these can be used very successfully as long as we do not forget that the only real basis for these formulas in a comparison of the calculated results with experimental data, and so they can be used only when a check gives positive results for the specific conditions. As more and more experimental data accumulate, these attempts can be extended to set up working formulas which are valid over wider ranges. However, in spite of its practical value, the "characteristic temperature" method must be regarded as purely empirical.

In summing up, we are forced to say that up to now there is no rigorous general method for taking the variability of the physical properties into account.

The dependence of the physical properties on temperature manifests itself in another interesting effect, known as the *effect of the heat flux direction*. If the physical properties are assumed to be constant, we can say quite validly that the direction of the heat flux has no effect on heat transfer, since, apart from the variability of the properties, we cannot show any way in which the process of heat transfer from the surface to the medium differs from that of heat transfer from the medium to the surface. Actually, at fairly small temperature driving forces (when the process can be regarded as approximately isothermal) the heat transfer intensity is the same in both cases, other factors being equal. However, the situation becomes more complicated when we consider the variability of the properties.

For reasons which will be clear from the preceding, it is difficult in the general case to predict completely the influence of the present effect. However, it seems indisputable that on changing from heating to cooling the intensity of heat transfer should vary in a continuous manner (this is equivalent to the statement that passage through the value $\Delta T = 0$ is not connected with a discontinuity).

Some interesting conclusions can be drawn for the special case of a gas. As we have explained, the effect of variability of the physical properties can be expressed directly by the temperature factor under gas flow conditions. If one of the temperatures given by the conditions (for example, the wall temperature) is denoted by T_1, and the other (the fluid temperature) by T_2, the temperature factor is written in the form $\frac{T_1}{T_2}$ $\left(\text{or } \frac{T_2}{T_1}\right)$. The physical constants must be referred to one of these temperatures. Let us agree always to refer them to the temperature in the numerator.

The direction of the heat flux is obviously determined by whether the temperature factor is a proper or improper fraction. If we adhere to our convention, the condition $\frac{T_1}{T_2} < 1$ corresponds to cooling, while $\frac{T_1}{T_2} > 1$ corresponds to heating. For given values of the wall temperature T_1 and the temperature driving force ΔT the heat transfer intensities under conditions of cooling and heating are obviously not the same, since these are really two different cases differing in the values of the temperature factor: in the first case this is equal to $\dfrac{T_1}{T_1 + \Delta T}$, and in the second, it is $\dfrac{T_1}{T_1 - \Delta T}$.

However, the problem can be set up so that the effect of the heat flux direction can be isolated. To do this it is necessary to compare the heat transfer intensities for the cases corresponding to the conditions $\left(\frac{T_1}{T_2}\right)' = \left(\frac{T_2}{T_1}\right)''$, where the single and double primes denote the quantities referring to the first and second cases respectively. Experiments show that under these conditions the heat transfer intensities are the same. This indicates that reversal of the process does not in itself have any effect on the heat transfer intensity. It must be remembered that all these discussions refer to flows of moderate velocities only. The direction of the heat flux plays an important part in high-speed flows.

Finally, we must note that the temperature dependence of the physical properties of the medium may also lead to more marked effects which are capable of changing the nature of the process (perturbation of a stream, superposition of free convection on forced flow). Under these conditions the fundamental equations have to be changed radically, since the usual corrections and additions do not suffice.

28. THE CASE OF HEAT FLUXES GIVEN BY THE CONDITIONS. THE DIMENSIONLESS FORM OF TEMPERATURE

In spite of their differences, all the problems considered so far have had certain common features. In all cases at least one temperature difference ΔT has been given by the conditions. However, problems can be set up in which the intensity of the thermal interaction between a surface and a fluid is defined directly by the conditions in terms of a *thermal loading* (a heat flux density q). This obviously means that the problem of the temperature field for the solid body has boundary conditions of the second type. Although it is less typical, this case is still of interest since it has to be investigated in the study of certain technically important processes (including boiler processes, heat transfer in nuclear reactors, etc.).

Under these conditions the generalized temperature distribution equations of the form obtained earlier cease to be valid, since the parametric value of the temperature (or more precisely, of the temperature difference ΔT) disappears from the conditions. We therefore have to derive an expression equivalent to ϑ_0 in order to construct a dimensionless temperature. From the equation

$$q = \lambda \left| \operatorname{grad}_n \vartheta \right|_0$$

we obtain the group

$$\pi = \frac{q_0 l}{\lambda \vartheta_0},$$

from which it follows directly that $\frac{q_0 l}{\lambda}$ is a quantity equivalent to

ϑ_0. In this case the dimensionless temperature is written in the

form $\frac{\vartheta}{\frac{q_0 l}{\lambda}} = \frac{\vartheta \lambda}{q_0 l}$. This expression must be introduced $\left(\text{in place of } \frac{\vartheta}{\vartheta_0}\right)$

as an unknown function in all the generalized equations for the

temperature.

Chapter IV

SIMILARITY

The concept of similarity is of great importance both in developing the fundamental ideas and mathematical apparatus of the method of generalized analysis and also in studying actual problems. Let us now consider this concept and develop certain arguments to strengthen the ideas given earlier.

A. Similar Transformations

29. THE SIMILARITY OF PHENOMENA AND SIMILAR TRANSFORMATIONS. TRANSFORMATION FACTORS FOR DIFFERENTIAL OPERATORS

Let us discuss first in its most general form the problem of the complete group of distinguishing features of an individual generalized case, or, in other words, *the conditions which are necessary and sufficient for phenomena to be similar.* It is necessary to return to this problem, since earlier we started from a definite form of the equation (p. 19), which, although it is very widely encountered, is nevertheless a special case. We will now get rid of this limitation.

Thus we are given an equation (or system of equations) with the corresponding conditions for uniqueness of solution. In order to be specific, it will be assumed that we are considering one equation containing one unknown variable. The solution of the problem is an equation which defines the unknown variable as a single-valued function of the independent variables and parameters. This solution corresponds to some specific (unique) phenomenon. Any other phenomenon similar to it is described by an equation which can be given by multiplying each of the quantities (the independent variables, parameters, and unknown variable) appearing in this solution by some factor which is a constant selected for these quantities. Different values of the factors correspond to different, but similar, phenomena. By giving the factors different values in succession, we obtain an infinite number of solutions which describe a group of similar phenomena. This operation is termed a *similar transformation.*

Thus, we are considering a whole collection of solutions, which determine a group of similar phenomena, as the result of a similar transformation of the basic solution with all the possible values of the transformation factors. (The basic solution corresponds to the phenomenon described by the conditions.) This argument follows from the very basis of the concept of similarity of phenomena. Briefly, it can be reduced to the clear and simple statement that similar phenomena are described by similarly transformed solutions. However, it is not immediately clear how these solutions are to be obtained. This is a very significant problem.

The solution of a properly stated problem is completely determined by its conditions, and the solution can only be affected by changing the conditions. Of course, any change in the conditions makes itself felt in a single-valued manner in the solution. However, it is extremely difficult (if not impossible) to establish the

exact nature of this relationship in the general case. In the mean-
while, we are faced with the problem (in the most general form) of
finding a method by means of which a definite change in the final
solution can be effected.

It is obvious that changes introduced in the conditions can have
no effect at all on the fundamental equations (since all mutually similar phenomena must belong to the same class). Consequently, only
the conditions for single-valuedness are subject to change, i.e., the
uniqueness conditions, when this concept is extended to include the
physical parameters of the system also. In addition, it must be
remembered that phenomena which are similar to one another differ
only in the numerical values of the quantities. Only the numerical
values of the given quantities need therefore be changed (but not
their distribution laws), as long as they are given by the conditions,
as in the case of the boundary-value problems which are our primary concern.

It is not difficult to conclude that these changes must represent
similar transformations. Suppose that there is some (second) phenomenon which is similar to a given (first) phenomenon. On comparing the two phenomena, we would observe that they differ only
by the numerical values of the quantities in such a way that the
value of any quantity for the second phenomenon is obtained from
the value of the same quantity for the first phenomenon by multiplying by some constant number, i.e., by means of a similar transformation. This relationship is satisfied in all cases, including the
parametric values given by the conditions.

Hence the numerical values contained in the conditions of the
second problem are obtained as a result of a similar transformation
of the corresponding values given by the conditions of the first
problem. In all other respects the two problems are identical
(obviously, similar transformations do not disturb the distribution

laws given by the conditions). This means that a similar transformation of all the parametric values with the fundamental equation unchanged corresponds to a change of the problem conditions, which leads to a similar transformation of the solution, i.e., to a similar phenomenon. This result can be stated briefly as follows: *in the absence of a change of the equation, a similar transformation of the conditions for single-valuedness leads to a similar transformation of the phenomenon itself.*

Thus, in order that phenomena be similar, it is necessary and sufficient that the following requirements be fulfilled: 1) the equations must be identical; 2) the conditions for single-valuedness must be similar. These are two independent requirements, and one is not simply complementary to the other. It is not clear whether they can be satisfied simultaneously, or whether they contradict each other in any way.

Even a simple consideration will show that there are serious foundations for these misgivings. In fact, the requirement that the equations remain unchanged when a similar transformation is performed on the variables means that the equation must be satisfied, first, by definite values of all the quantities (the first solution), and, second, by the same values multiplied by an arbitrary constant number (the second solution). It is quite obvious that in the general form it is absolutely impossible to fulfill this requirement. The equation must therefore possess some special properties (this, generally speaking, is inconsistent with our attempt not to limit ourselves by any assumptions as to the structure of the equations), or the transformation factor must be chosen in some special way. In any case, the problem of the simultaneous fulfillment of both requirements (identity of the equations and similarity of the conditions for a single-valued solution) requires special consideration. In this way, we have arrived at *the problem of the conditions for invariance of equations with respect to similar transformations.*

Before passing on to the study of this problem, we must stop to consider a purely technical procedure. We will have to deal with various differential equations. It is therefore necessary to establish the rules for a similar transformation when the variable to be transformed appears under the sign of a differential operator.

Suppose that the variable z is given in terms of the variables x and y in the form

$$z = \frac{dy}{dx}.$$

In this case a similar transformation of the variables leads to the equation

$$k_z z = \frac{d\,(k_y y)}{d\,(k_x x)},$$

where k denotes the transformation factors, and the subscripts refer to the quantity being transformed. As constants, the factors k can be removed from the differentiation sign, to give

$$k_z z = \frac{k_y}{k_x}\frac{dy}{dx} = \frac{k_y}{k_x} z.$$

Hence

$$k_z = \frac{k_y}{k_x}.$$

For a second derivative

$$u = \frac{d^2 y}{dx^2} = \frac{dz}{dx};$$

and on the basis of the preceding case, we have

$$k_u = \frac{k_z}{k_x}$$

or

$$k_u = \frac{k_y}{k_x^2}.$$

In an exactly similar way, the problem of the transformation factor for a third-order derivative can be reduced to that of finding the factor for a second-order derivative, or, more generally, the problem of a derivative of order m can be reduced to that of a derivative of order $(m-1)$. Thus, if

$$v = \frac{d^m y}{dx^m},$$

then

$$k_v = \frac{k_y}{k_x^m}.$$

However,

$$k_v = K_{\left(\frac{d^m y}{dx^m}\right)} \quad \text{and} \quad \frac{k_y}{k_x^m} = K_{\left(\frac{y}{x^m}\right)},$$

where the new symbols require no explanation.

Thus

$$K_{\left(\frac{d^m y}{dx^m}\right)} = K_{\left(\frac{y}{x^m}\right)}.$$

It is highly significant that the expression $\frac{y}{x^m}$ is the reduced group referring to the derivative $\frac{d^m y}{dx^m}$ as the simplest differential operator. Hence, using the symbols established earlier,

$$K_D = K_\Pi. \qquad (4.1)$$

It is easy to show that this result, which has been obtained for the simplest form of operator, remains valid for any homogeneous operator, regardless of its complexity. Actually, the transformation factors for the individual terms appearing within a homogeneous operator can differ only with respect to the subscripts corresponding to the different components of the vector. However, the factors for a similar transformation of the individual components obviously must be equal to each other and also equal to the transformation factor for the vector as a whole (this means that in general it is unnecessary to write the subscripts for the components in the transformation factors). In this case, all the components have the same transformation factor, and, as a result, this is also the transformation factor for the operator as a whole (K_D). It is quite clear that the quantity K_D is none other than the transformation factor for the reduced group II corresponding to the operator D. We can draw the following general conclusion: *the transformation factor for a homogeneous operator is defined as the transformation factor for the corresponding reduced group.* From this it follows in turn that, in studying the properties of similar transformations, we can consider the corresponding reduced groups instead of the operators. This means that in general it is unnecessary to investigate the fundamental equations with all their very real complexities. They are replaced by the simple expressions which are obtained by replacing the operators by the power groups. It is easy to see what a great advantage it is to be able to reduce the whole investigation to a consideration of the equations written in their final form.

30. THE INVARIANCE OF EQUATIONS WITH RESPECT TO SIMILAR TRANSFORMATIONS. HOMOGENEITY. ABSOLUTELY AND CONDITIONALLY HOMOGENEOUS FUNCTIONS. THE NUMBER OF DEGREES OF FREEDOM

Let us consider the equation

$$F(x_1, x_2, ..., x_n) = 0, \tag{4.2}$$

where F denotes a function of arbitrary type, and $x_1, x_2, \ldots, x_n$ are the quantities of importance for the problem (independent variables, unknown variables, parameters).

Now let us carry out a similar transformation of all the quantities (i.e., let us replace each of the quantities x_i by the product $k_i x_i$, where k is a constant number). If the equation is to remain unchanged by this, we find from (4.2) that

$$F(k_1 x_1, k_2 x_2, \ldots, k_n x_n) = 0. \qquad (4.2')$$

Both equations contain the very same values of the quantities x_i and so they must be considered together as a single system of equations.

Thus our problem is reduced to that of the conditions of compatibility of equations (4.2) and (4.2'). Naturally these equations do not express any relationship between the quantities x_i and the constant factors k_i. The transformation factors can be interrelated, but they cannot be defined in terms of the quantities being transformed. In this case, there is only one method to satisfy both equations simultaneously. It is necessary for the expression $F(k_1 x_1, k_2 x_2, \ldots, k_n x_n)$ to tend identically to zero when the quantities x_i assume the values which satisfy Eq. (4.2). This is possible only if

$$F(k_1 x_1, k_2 x_2, \ldots, k_n x_x) = \varphi(k_1, k_2, \ldots, k_n) F(x_1, x_2, \ldots, x_n) \qquad (4.3)$$

or, in contracted form,

$$F[k_i x_i] = \varphi[k_i] F[x_i]. \qquad (4.3')$$

Equation (4.3) indicates that, when all the quantities appearing within the function are multiplied by constant factors, all the factors can be taken outside the function sign, and form some new factor for the function as a whole. Thus the function must possess the special property that a similar transformation of the individual variables leads to a similar transformation of the function as a whole. This remarkable property is known as homogeneity, and functions possessing it are said to be homogeneous. Thus the condition necessary and sufficient for invariance of an equation with respect to a similar transformation is that the function comprising the left-hand part of the equation must be homogeneous. This condition is a very drastic limitation, since the property of homogeneity is characteristic of only one very narrow class of functions. Let us prove this.

Let f be a homogeneous function of the variables x_1, x_2, ... , x_n. In the present case we will use the notation $x_i' \equiv k_i x_i$, which enables us to write

$$f(x_1', x_2', ... , x_n') = \varphi(k_1, k_2, ... , k_n) f(x_1, x_2, ... , x_n),$$

or, in abbreviated form,

$$f[x_i'] = \varphi[k_i] f[x_i],$$

where the symbol [ ] denotes, as before, "of all x_i."

Bearing in mind that all the x_i' depend on the corresponding k_i, let us differentiate both parts of the equation with respect to one of the k_i, for example, with respect to k_1. This gives

$$\frac{\partial f[x_i']}{\partial k_1} = \frac{\partial \varphi}{\partial k_1} f[x_i].$$

However,

$$\frac{\partial f[x_i']}{\partial k_1} = \frac{\partial f[x_i']}{\partial x_1'} \frac{\partial x_1'}{\partial k_1} = \frac{\partial f[x_1']}{\partial x_1'} x_1.$$

Therefore,

$$\frac{\partial f[x_i']}{\partial x_1'} x_1 = \frac{\partial \varphi}{\partial k_1} f[x_i].$$

All these relationships are valid for any values of the factors k_i. Let us assume

$$k_1 = k_2 = ... = k_n = 1.$$

Hence

$$x_i' = x_i$$
$$i = 1, 2, \dots n.$$

The previous equation then assumes the form

$$x_1 \frac{\partial f[x_i]}{\partial x_1} = f[x_i] \left| \frac{\partial \varphi}{\partial k_1} \right|_{[k_i]=1}.$$

But

$$\left| \frac{\partial \varphi}{\partial k_1} \right|_{[k_i]} = a_1,$$

where a_1 is some constant. Substituting this value in the previous equation, we find

$$x_1 \frac{\partial f}{\partial x_1} = a_1 f$$

or

$$\frac{1}{f} \frac{\partial f}{\partial x_1} = \frac{a_1}{x_1}.$$

The function f is determined by this equation in the form

$$f = C_1 x_1^{a_1}.$$

Here C_1 is a quantity which depends on all the remaining variables except x_1.

By repeating this treatment for all the variables in turn, we finally find

$$f = C_1 x_1^{a_1} x_2^{a_2} \dots x_n^{a_n}, \tag{4.4}$$

where C is a constant.

Thus the function possesses the property of homogeneity only when it can be represented by a power group. It follows from this that the property of invariance with respect to a similar transformation is possessed by equations whose left-hand parts represent homogeneous operators (e.g., the continuity equation for incompressible fluids). Of course this solution cannot possibly satisfy

us. It so limits the possibilities of analysis that it becomes practically totally useless, and it is necessary to find other routes.

Up to now we have considered the question of the invariance of
equations in the most general form, without limiting the freedom
of the similar transformation by any special assumptions. It has
been shown that under these conditions the problem becomes impossible, since it reduces practically to the problem of the limitations to which the structure of the equation must be subjected in
order to possess invariance with respect to any similar transformation. This is a very stringent limitation. Let us therefore set
up the opposite problem. We will attempt to solve the problem of
the limitations with which the similar transformations must comply
in order that the equation will be invariant, regardless of its
structure. The concept of this solution is very simple.

It is quite clear from Eqs. (4.2) and (4.2´) that the invariance
of the equation is unconditionally guaranteed by the requirement
$k_1 = k_2 = ... = k_n = 1$. Of course, this is a quite trivial case without
particular interest, since it corresponds not only to identity of the
equations but also identity of all the quantities appearing in the equations. However, if an analogous operation were possible with respect
to any other variable, we would obtain a result which would be not at
all trivial. This method of solution can be easily put into practice.

Let us introduce power groups as the new variables. When we
carry out a similar transformation of the quantity x_i the new variable P_j is also transformed similarly as a homogeneous function.
The transformation factor for the group as a whole, K_j, is constructed from the transformation factors for the individual variables, k_i, in just the same way that the group P_j is formed from
the variables x_i. Actually, if

$$P[x_i] = Cx_1^{a_1}x_2^{a_2} ... x_n^{a_n},$$

then

$$P[k_i x_i] = C\left(k_1^{a_1} k_2^{a_2} \dots k_n^{a_n}\right) x_1^{a_1} x_2^{a_2} \dots x_n^{a_n}$$

and so

$$K = k_1^{a_1} k_2^{a_2} \dots k_n^{a_n}.$$

If it is now required that all the factors K_j should tend to one, the equation proves to be invariant, but the solution is not trivial. Thus, returning to Eq. (4.2),

$$F(x_1, x_2, \dots, x_n) = 0.$$

Each of the power groups contained in the function is used as a new variable. Their common part, denoted as P_0, is removed from inside the function sign. This gives

$$F(x_1, x_2, \dots, x_n) \equiv P_0 \Phi(P_1, P_2, \dots, P_r). \tag{4.5}$$

Naturally, in the general case the number of new variables (r) is not equal to the number of initial variables (n). Equation (4.2) can now be broken down into two independent equations:

$$P_0 = 0$$

and

$$\Phi(P_1, P_2, \dots, P_r) = 0.$$

The first equation is not of interest, since it returns us to the elementary case of a homogeneous function (a homogeneous operator) which was considered earlier. On the other hand, the second equation leads to a new result. After a similar transformation of the quantity x_i it assumes the form

$$P|K_j P_j| = 0.$$

All the factors K_j must become equal to unity for the equation to be invariant. It is obvious that this is a necessary and sufficient condition to cancel out all the changes in structure which arise in the transformation of the quantities x_i. Thus, let us assume:

$$K_j = 1.$$
$$j = 1, 2, \dots, r.$$

(4.6)

The factors K_j represent power groups formed from the transformation factors k_i. Thus each of the equations (4.6) establishes a relationship between the factors k_i. This means that the system of equations (4.6) should be regarded as an expression of the whole collection of limitations which govern the choice of the factors k_i for a similar transformation.

Let us summarize the results obtained above. For an equation to be invariant with respect to similar transformations of the variables appearing in it, it is necessary and sufficient that the function comprising its left part be homogeneous. Functions exist which are homogeneous in structure. The homogeneity of these functions does not depend on any additional assumptions as to the properties of the transformation. These functions are logically termed absolutely homogeneous. Only power groups have the property of unconditional homogeneity. For any other functions to be homogeneous, the transformations must comply with special requirements. These requirements can be expressed in the form of Eq. (4.6), which gives an interrelationship between the transformation factors, and so limits the freedom of their choice to some extent (depending on the structure of the function). In contrast to the absolute homogeneity of the power functions, the homogeneity of this type depends entirely on additional rules established for the transformation. Thus in the general case the transformation is limited by special requirements by means of which a function which is not homogeneous in itself is artificially converted into a conditionally homogeneous function.

It follows from all this that the system of equations (4.6) expresses a complete block of conditions; it is necessary and sufficient to satisfy these for the equation to be invariant with respect to similar transformations. Let us consider these conditions more closely. We have a system of r equations with respect to the unknowns k_i, the total number of which is n. This means that it is possible to choose arbitrarily $(n - r)$ values of the factors k_i (we will not consider special cases which can arise when there are special relationships between the exponents). It is customary to term the difference $(n - r)$ *the number of degrees of freedom* of the transformation.

As long as n is one larger than r, there will be at least one degree of freedom, and hence, by giving one of the factors an arbitrary value, we can obtain an infinite number of special solutions.

If $n = r$, only one solution exists, and no freedom remains in choosing values of the factors k_i. This solution can be obtained directly, since it is immediately obvious that the system (4.6) is

satisfied if all the k_i values are chosen as unity. The trivial solution $k_1 = k_2 = \cdots k_n = 1$, which is valid under all circumstances regardless of the relationship between r and n, means that the transformation cannot be performed (and hence that only the one primary phenomenon is being considered). Thus a transformation is possible only when $n > r$. This requirement is satisfied in the overwhelming majority of cases.

31. THE NECESSARY AND SUFFICIENT CONDITIONS FOR SIMILARITY

Thus, similarity of phenomena is governed by definite qualitative requirements, which can be given in the form of the system of equations (4.6). It is not difficult to understand the significance of the requirements expressed in these equations. The condition $K_j = 1$ implies the invariance of P_j. If all the K_j values tend to unity, all the P_j are invariant. Consequently, the system (4.6) is none other

than a collection of conditions which ensure that all the power groups
are unchangeable and that the equations are invariant when the
quantities x_i are transformed similarly. In other words, each group
must have the same value for all similar phenomena, and this is
the only quantitative requirement that limits the freedom of the
transformation of a parametric value of a quantity given directly
by the initial conditions of the problem on passing to the new conditions (which define phenomena similar to the first phenomenon).
This means that *the **only** condition necessary (and sufficient) for
similarity of the phenomena is equality of the values of the groups
made up of the quantities given in the conditions* (this corresponds
to similarly transformed conditions, i.e., conditions which are
obtained from one another by transforming similarly all the quantities contained in the conditions). It is obvious that these values
are the same for all mutually similar phenomena, and are quantitative characteristics of each given group of similar phenomena—
for each generalized case.

The problem of the conditions necessary and sufficient for phenomena to be similar has been solved with very general assumptions.
It is easy to see that the general solution includes the solution obtained earlier as a special case corresponding to equations of the
type (2.1). Actually, if the left-hand part of the equation can be
represented in the form of a sum of homogeneous operators, then
one of the operators, which we can denote by D_0, can be removed
from the parentheses, after which we pass automatically to the
relative operator

$$d_{j0} \equiv \frac{D_j}{D_0} :$$
$$D_0 + D_1 + D_2 + \cdots + D_r = D_0 (1 + d_{10} + d_{20} + \cdots + d_{r0}).$$

It is obvious that this expression reduces to the form $P_0 \Phi (P_1,
P_2, \ldots, P_r)$ which we have already found from Eq. (4.5). The reduced

group Π_0 (defined by the operation $D_0 \to \Pi_0$) corresponds to P_0 and the reduced group π_{j0} (for which, by analogy, $d_{j0} \to \pi_{j0}$) corresponds to $\dfrac{P_j}{P_0} \equiv p_{j0}$. Thus, in complete agreement with our usual arrangement, the reduced groups have the significance of fundamental quantitative characteristics of similar phenomena. Their importance as "similarity groups" is also completely verified.

In the general case (when the left-hand part of the equation is not a sum of homogeneous operators and it is impossible to separate a factor which is common to all the terms) all the results above remain valid when the appropriate changes are made. The main difference is that it is necessary to consider the operators themselves (directly in the form in which they appear in the equation) rather than their ratios. Thus the Π_j assume the role of the fundamental quantitative characteristics which determine the properties of the similar phenomena; under these conditions these become the similarity groups. It is very significant that under these conditions also the similarity groups are dimensionless quantities (though this is not immediately clear, since the Π_j, in contrast to the π_{j0}, are not relative quantities). Thus for a polytropic process, described by the equation $\dfrac{p}{\rho^n} = \text{const}$ (p is the pressure, ρ is the density), the polytropic exponent n is obviously a similarity group. This quantity is dimensionless under all conditions (a pure number), though it is not relative.* We will show later more rigorously (p. 234) that the dimensionless nature of the groups Π_j arises from their invariance with respect to similar transformations (i.e., from a fundamental property of these quantities which is one of their characteristics as similarity groups).

---

*However, by taking logarithms and differentiating both sides of the equation $p = \text{const } \rho^n$, we have $\dfrac{dp}{p} = n\dfrac{d\rho}{\rho}$, whence $n = \dfrac{dp}{p} \Big/ \dfrac{d\rho}{\rho}$. The exponent n is therefore obtained as a relative quantity.

B. The Method of Models

32. ESSENTIAL FEATURES OF THE METHOD OF MODELS. THE ARRANGEMENT OF EXPERIMENTS AND LIMITING CONDITIONS

One of the main features of generalized analysis as a method of investigation is that each phenomenon is considered as a representative of a whole multitude of phenomena similar to it (all the phenomena contained in one generalized case). In the ideas of generalized analysis, the investigation of some specific phenomenon essentially involves the study of the properties of the group of similar phenomena as a whole. Thus a knowledge of the properties of some phenomenon which can be studied serves as the basis for investigating any other phenomenon similar to it. In other words, a phenomenon to be studied can be replaced as the object of investigation by any arbitrarily chosen phenomenon which is similar to it. A very useful and widely used method of experimental investigation is based on this principle; it is known as the method of models.

It follows from the foregoing that the method of models consists in replacing the object of the experiment. In order to obtain a correct estimate of the importance of this method, let us turn our attention to the following discussion. If an investigation is being carried out to explain the general laws of a process, the experimenter has wide possibilities for choosing the general arrangement of the experiments and in setting up conditions favorable for carrying them out. In this case, the theory of similarity is a special method of investigation (which can be used in the preliminary analysis of the problem, when the system of experiments is being developed, and in the final treatment of the experimental results). However, in addition to these investigations, which are aimed at the solution of general problems of a fundamental nature, other investigations of a quite different type are also of great importance, particularly in engineering practice. Quite frequently it is necessary to study in detail some quite specific process, which develops into a system with definite geometric and physical properties and with given regime conditions.

A characteristic example of this type of problem is the experimental investigation of the operation of a definite piece of equipment. It is obvious that in this case the phenomenon to be studied is given with the greatest possible physical precision. The object of the experiment is fixed as a unique, specific phenomenon. Under these conditions the experimenter completely loses his freedom of choice: the experimental arrangement is predetermined in all respects, down to the values of the regime parameters. This ultimate rigidity of the requirements for the arrangements for the experimental investigation quite often leads to extremely important difficulties which are quite unresolvable in practice. The usefulness of the method of models makes itself felt very forcibly in this situation. The replacement of the given phenomenon by another phenomenon— the model—frequently makes it possible to solve all the experimental difficulties: the only requirement for doing this is that the model must be similar to the original phenomenon.

It is easy to see what opportunities are opened by the possibility of choosing the dimensions of the system (but, of course, not its configuration), the rate of development of the process, the physical medium, and the values of the regime parameters (but not the boundary distribution laws). However, it would be wrong to suppose that the change from the original phenomenon to the model makes it possible to select arbitrarily all the quantitative experimental conditions. The fundamental requirement—similarity between the original phenomenon and the model—is a source of important limitations.

An investigation carried out by the method of models includes two essentially independent experimental problems: it is first necessary to reproduce a phenomenon similar to the original, and then it is necessary to carry out the required observations and measurements on it. We are most interested in the first part of the program, since the specific features of the method of models occur in this part. The problem of how to establish experimentally a phenomenon which is similar to that under consideration (which is defined as a phenomenon corresponding to a specific numerical variant) leads us back again to the well-known problem of the conditions necessary and sufficient for the similarity of two phenomena. We have considered this problem in great detail, and without further discussion it is clear that the free choice of the values of the parameters which can be made in the experiments is limited by the system of equations (4.6):

$$K_1 = K_2 = \cdots = K_r = 1.$$

This limitation can be expressed in another way as the requirement that all the similarity groups characterizing the properties of the model should have specified values (which are derived from the conditions defining the properties of the original phenomenon). From this it follows that in the case of a group of parameters totaling n there are $(n - r)$ degrees of freedom (i.e., $n - r$ parameters can be specified arbitrarily).

It is clear that the limiting equations can contain only the quantities (and, correspondingly, their transformation factors) for which parametric values are given by the conditions. If a parametric value of some variable is not given, its transformation factor is not selected, but determined in terms of the transformation factors of the other quantities which form a group equivalent to it.

Thus the real advantage in going from the sample to the model depends essentially on the relationship between the total number of parameters and the number of limiting equations. Let us consider a simple example by way of illustration.

Suppose that the method of models is being applied to the investigation of the steady-state motion of an incompressible fluid (liquid or gas), in which experiments are carried out with the object of studying in detail the kinematics of the situation and the dynamic interaction of the stream with a solid body. This type of problem is very characteristic of the conditions which arise in the design of a new equipment, when it is impossible to check on the properties of the equipment (which may still not have progressed beyond the planning stage) by comparing the calculated results with experimental data. If we denote the quantities referring to the actual case by a single prime, and those referring to the model by a double prime, the equations which limit the free choice of the model parameters can be written in the form:

$$\mathrm{Re}'' = \mathrm{Re}' \text{ and } \mathrm{Fr}'' = \mathrm{Fr}'$$

or

$$\frac{k_w k_l}{k_\nu} = 1 \text{ and } \frac{k_w^2}{k_g k_l} = 1.$$

Of course, the accelerating force of gravity remains the same in both cases, and so we can put $g'' = g'$ (or $k_g = 1$). Thus, three parameters (w_0, l, and ν) are related by two limiting conditions. This means that the experimenter has only one degree of freedom in setting up the model. For example, if the working fluid is selected (i.e., if the value of the factor k_ν is fixed), all the parameters of the model are fixed by this (since the transformation factors for the velocity and dimension are defined directly by the equations $k_w = \sqrt[3]{k_\nu}$ and $k_l = \sqrt[3]{k_\nu^2}$). In connection with this it is interesting to note that this problem generally yields to modeling only when different fluids are used in the model and the actual case. Otherwise (i.e., with $k_\nu = 1$) the number of degrees of freedom becomes equal to zero, and the possibility of modeling is completely lost, as long as we are not thinking of going to extraordinary means, such as varying the value of g. A more favorable situation arises in the case of pure forced motion, since there is only one limiting requirement under these conditions:

$$\mathrm{Re}'' = \mathrm{Re}', \text{ or } \frac{k_w k_l}{k_v} = 1.$$

It is obvious that now, with two degrees of freedom, the experimenter does not have to exclude the possibility of having $k_v = 1$ for fear of making modeling impossible. It must be noted that the use of the natural fluid in the model is frequently a great technical advantage. However, this leads to a very troublesome complication, for, since $k_v = 1$, the transformation factors for the velocity and dimension are rigidly interrelated by the condition $k_w k_l = 1$, and this leads to two opposing tendencies. Thus, in carrying out model tests of a rapidly moving body of large dimensions (for example, an aircraft) it would be very desirable to carry out the experiments on a scaled-down model and at reduced velocities, primarily because of the reduced energy consumption of the air stream. Howevery, reducing the dimensions has as a necessary consequence a corresponding increase (rather than the desired decrease) in the velocity, while decreasing the velocity requires the use of an enlarged (rather than reduced) model. The two requirements are therefore clearly incompatible. A radical solution to this problem is to back off from the condition $k_v = 1$ (for instance, to use as the working fluid air at increased pressure, i.e., at reduced kinematic viscosity).

If we are considering an unsteady-state process (or studying the motion of a particular particle), another important quantity must be borne in mind—the duration of the process—and as a result there is still another limiting condition (p. 113):

$$\frac{k_w k_t}{k_l} = 1.$$

The simultaneous addition of one more quantity and one more limiting condition cannot lead to any change in the number of degrees of freedom. In formal terms, the new relationship determines the transformation factor of the new quantity, and everything else remains unchanged. However, in fact the role of the homochronicity group in the system of relationships defining the experimental conditions (for arranging the experiments by the method of models) is considerably more important than this. The choice of a value for the time transformation factor provides a means for directly influencing the rate of development of the process which is being studied. It is scarcely necessary to mention the advantages which accrue from the corresponding changes in the rate of development of the process.

It will be sufficient to mention the case of a very rapidly occurring process which can be greatly slowed down by means of modeling, as a result of which it becomes possible to carry out detailed visual observations and to perform all the necessary measurements with ease. Thus, the factor k_t is correctly placed among those which can be chosen arbitrarily. If we assume that the choice of the factor k_v (choice of working fluid) is made arbitrarily also, this consumes all the degrees of freedom. The remaining transformation factors, which determine the dimensions of the model and the fluid velocity, are obtained from the equations

$$k_l = \sqrt{k_v k_t} \quad \text{and} \quad k_w = \sqrt{\frac{k_v}{k_t}} \, .$$

In connection with our discussion of the role of the homochronicity group, let us consider the problem of modeling a cooling (or heating) process in a solid body as one of the characteristic cases in which the method of models is applied to the study of unsteady-state processes. Assuming that the problem has been set up with boundary conditions of the third type, we have one limiting equation:

$$\mathrm{Bi}'' = \mathrm{Bi}', \text{ or } \frac{k_a k_l}{k_\lambda} = 1.$$

Thus, in passing from the sample to the model, the intensity of heat transfer must be changed in correspondence with the changes in the system dimensions and physical properties of the media. Thus, in carrying out the experiments, the experimenter must have the means for realizing a definite, previously selected heat transfer intensity. This requirement cannot always be realized, since in many cases the experimenter does not have sufficient information about the laws which determine the intensity of heat transfer between a body and a medium surrounding it. It must be admitted that favorable arrangements for the efficient application of the method to the study of unsteady-state temperature fields in solid bodies occur practically only in the case when the problem can be set up with boundary conditions of the first type (in particular, when boundary conditions of the third type degenerate to those of the first type due to the tendency $\mathrm{Bi} \rightarrow \infty$). With this assumption all the difficulties disappear directly. The limiting condition (unique!) loses its validity, and we have a typical case of self-similarity. All the parameters of the model can now be chosen quite arbitrarily. Similarity of the phenomena is guaranteed by the geometric similarity of the system and similarity of the boundary distributions of the temperature (of the temperature difference, to be exact). It is only necessary to remember that in studying the process under these conditions (monotonic temperature changes) the dimensionless form of the time appears in the form $\frac{at}{l^2}$ (since it is impossible to establish a characteristic value for the time). This means that the time transformation factor cannot be chosen arbitrarily, but is determined by the choice of the modeling material (k_a) and the geometric scale of the model (k_l) according to the equation

$$\frac{k_a k_t}{k_l^3} = 1.$$

We should note the great dependence of the rate of the process on the dimensions of the system. For example, if the model is made 1/10th of the natural size, then (assuming $k_a = 1$) the process will occur 100 times more rapidly in the model than in the sample. It is easy to see the practical importance of this result if we bear in mind that many processes involving the rearrangement of temperature fields occur extremely slowly in nature (thus, in the case of heat conduction in the soil, temperature distributions which can practically be regarded as steady-state are set up in the course of several months).

Now let us attempt to estimate the possibilities of the method of models as applied to processes which involve both fluid motion and heat transfer. Compared to the problems considered above, this more complicated case differs by one new limiting condition:

$$\mathrm{Pr}'' = \mathrm{Pr}', \text{ or } \frac{k_v}{k_a} = 1.$$

The addition of this requirement greatly complicates the situation, since it is extremely difficult (practically impossible) to select a modeling fluid which is suitable for use and at the same time not too different with respect to the value of Pr. For instance, water, which is the classical working fluid in modeling gas flows, differs greatly from gases in its value of Pr. A more or less suitable value for Pr (around unity) is obtained at temperatures above 150°C (and therefore at pressures greater than 5 absolute atmospheres). Of course, with these parameters it is impossible to regard water as a suitable working fluid. Actually, the requirement that the values of the Prandtl group should be equal practically prohibits the use of a different modeling fluid. However, as we have explained, the use of a different fluid in the model is an important means of improving the experimental conditions. From this it appears that it would be possible to reduce the number of degrees of freedom mentioned above. This would lead to considerable additional difficulties, and under certain conditions it would be quite impossible to set up a model.

To sum up, we must note that to include among the circle of problems being investigated processes which involve the transfer of heat in moving fluids leads to the addition of only one new limiting condition. However, this single additional condition is the source of great complications, since it limits the freedom of transforming not the new variable (the temperature), but a new physical constant of the fluid (the thermal conductivity), and hence limits the freedom of choosing the working fluid itself. Besides, it must be noted that the presence of heat transfer leads to additional difficulties in modeling apart from the requirement that the values of Pr for the two fluids—the natural fluid and that used in the model—should be the same. Strictly speaking, under these conditions it is quite impossible to substitute a liquid for a gas (or vice versa), since the volume of a gas varies in direct proportion to its absolute temperature, while a liquid provides a medium whose volume is almost constant. This difference becomes apparent for any appreciable change in temperature.

33. APPROXIMATE MODELING. THE DEGENERATION OF SIMILARITY GROUPS AS AN EFFECT WHICH INCREASES GRADUALLY. APPROXIMATE SIMILARITY. THE DEGREE OF DISTORTION DUE TO THE APPROXIMATION

A general conclusion which is becoming more and more clear the further we extend the range of these applications of the method of models is that the limiting conditions, which guarantee that the phenomena are similar, turn into sources of difficulties when we try to satisfy them—these difficulties are often quite serious, and sometimes quite unresolvable. This leads us to consider the question of whether it is actually necessary (not only in principle, but in practice) to see that all these conditions are satisfied exactly. In other words, we have to see to what degree all the limiting requirements are essential under actual given conditions, and to see how the nonfulfillment of various conditions affects the results of the investigation. This approach to the problem is suggested by our general concept of the similarity groups as quantitative characteristics of the process properties. The very same group may affect the development of processes to quite different degrees depending on the actual circumstances which result.

The familiar effect of group degeneration indicates that groups which under some conditions (at definite values of the parameters) have an all-important effect on the course of the process completely lose this effect under other conditions (at other values of the parameters). A detailed discussion of this effect has shown that there is no clearly marked boundary to the regions in which one or another of the groups degenerates. The degeneration of a group is correctly regarded as a quite gradual process in which the effect of the group decreases gradually.

All these features of the behavior of the groups make themselves felt correspondingly in the generalized equations. All the similarity groups which are of importance for a process are arguments of the generalized equations. However, the effect that each of them has varies as the process parameters change, and in the region of their degeneration the groups generally disappear from among the arguments.

The experimenter who wishes to set up an investigation correctly on the basis of the theory of similarity will undoubtedly become aware of evidence in these arguments that a freer approach to the problem of modeling is possible. It is obvious that it would be incorrect to regard the limiting conditions as a group of unalterable laws whose fulfillment is strictly obligatory in setting up the model (since any deviation from them infringes similarity, and so the results obtained lose their value as the basis for our ideas about the properties of the sample). It is more correct to regard these conditions as the maximum requirements which will completely guarantee the exact similarity of the phenomena, and hence to regard them as indisputably sufficient, but not entirely necessary, for the experiment to have the required cognitive value.

We can see quite clearly that in the region in which a group degenerates, the corresponding limiting equation must disappear. However, how should the region of degeneration be determined in practice? Here we will start not from a theoretical argument (as a similarity group increases or decreases without limit it ceases to be important in the process), but from the ideas that we have just assumed (as a group degenerates, its effect on the process gradually decreases). Thus the characteristic on which the degeneration of groups is actually based is that its change should not have any effect on the experimental results (or on a numerical solution). However, this is a very conventional characteristic, and arguments based on it depend on the accuracy of the experimental (or numerical) results. We must therefore draw the conclusion that ultimately a group must be regarded as degenerate or nondegenerate depending entirely on some agreed requirement of accuracy.

These ideas are of great importance. Together with strictly defined and exact theoretical ideas of the similarity of groups, it is also possible to have new concepts which are closer to the spirit of an experimental investigation. According to these new concepts, similarity and dissimilarity do not stand opposed to each other as exact opposites, distinguished by definitely contradictory properties. The opposition of similarity to dissimilarity assumes a relative nature. Dissimilarity passes into similarity (and similarity into dissimilarity) through an intermediate form, which can be correctly defined as *approximate similarity.* It is not difficult to give this concept a definite quantitative form.

It is assumed that we are comparing two phenomena with respect to their fields of the variable u (u' for the first phenomenon, u'' for the second). Under conditions of exact similarity we must have $u'' = ku'$ for any point in the field (i.e., at analogous points of the systems being compared), where k is a constant number which has been defined in advance (before setting up the experiment), in terms of the parametric values of u, so that $k = \dfrac{u_0''}{u_0'}$. (If u_0 is not given by the conditions, then the definition is in terms of the ratio of equivalent quantities.) Let us now assume that in setting up the experiment the group of limiting conditions is not completely satisfied, so that the phenomenon which is reproduced cannot be regarded as an exact model. In this case $u'' \neq ku'$.

Let us assume that

$$u'' = ku' + \Delta u'', \text{ or } \Delta u'' = u'' - ku'.$$

The quantity $\Delta u''$ is a quantitative measure of the degree of dissimilarity. Obviously it represents the error which we would make in determining u'' with the assumption that the phenomena were exactly similar. In addition to this absolute measure, let us also introduce a relative measure of the dissimilarity:

$$\varepsilon = \frac{\Delta u''}{u''} = 1 - k\frac{u'}{u''}, \tag{4.7}$$

which is termed the *degree of distortion*. Naturally, ε (like $\Delta u''$) varies from point to point and, under unsteady-state conditions, also varies with time. We will always think of the largest value. It is quite obvious that whenever the degree of distortion does not exceed the error in the results of the experimental comparison of u'' and u', it is impossible to observe any deviation from exact similarity. Thus as long as the degree of distortion falls within the limits of the experimental error, approximate similarity is identical in practice to exact similarity. What we actually wish to say is that the unused limiting conditions are not essential for the experiment, and in practice it is impossible to give any other significance to the concept of group degeneration.

Let us assume that the model is characterized by a degree of distortion ε. In this case the error of the results referring to the original problem is determined as follows. If the value obtained directly in the experiment is u'', the corresponding value of u' is determined according to Eq. (4.7) in the form

$$u' = u''\frac{1-\varepsilon}{k}.$$

On the other hand, the calculated value (denoted by u'_1), which can be regarded as correct, is given by the simple product $\frac{1}{k}\,u''$. Thus the absolute error caused by the approximate nature of the modeling will be

$$\Delta u' = u'_1 - u' = \frac{\varepsilon}{k}\,u''.$$

Similarly, the relative error is found to be

$$\frac{\Delta u'}{u'} = \frac{\varepsilon}{1-\varepsilon}.$$

If ε is regarded as a small quantity, this result can be rewritten as:

$$\frac{\Delta u''}{u'} = \varepsilon + \varepsilon^2 + \dots$$

We see that with an accuracy of quantities up to ε^2 the relative error in determining the quantities referring to the sample is equal to the degree of distortion which characterizes the model. Thus, knowing the degree of distortion, we can estimate quite fully the quality of the results obtained under conditions of approximate modeling. In particular, if ε does not exceed the experimental error, it is undoubtedly quite valid to apply approximate modeling. Unfortunately, in most cases it is impossible to determine the value of ε in advance without carrying out special experiments. We must remember that ε depends markedly on the special conditions of the process. Thus, for example, in the flow of a gas within a straight tube, compressibility does not often become important (and so the group $M*$ degenerates) for all cases up to a value $M = 0.7$. In the case of motion along curvilinear channels (in the spaces between the blades of a turbine) compressibility begins to have more of an effect on the detailed picture of the process (for example, on the pressure distribution) even at $M = 0.3$, and at $M = 0.6$ compressibility begins to influence even such overall effects as the hydrodynamic resistance.

As an illustration, let us consider the following example. The flow of an incompressible fluid within a tube is to be modeled. The motion of particular particles is being studied; for example, a small amount of dye is introduced at some point in the stream, and the manner in which it gradually spreads is observed (the dye must be chosen so that its specific gravity is the same as that of the liquid). The same liquid is used in the actual problem and the model. In the general case this process is characterized by two degrees of freedom (three quantities related by one condition,

---

*The group M (the Mach number) is the ratio of the velocity of the medium to the velocity of the propagation of sound in it; it characterizes the compressibility of fluids.

$\dfrac{k_w k_t}{k_v} = 1\Big)$. One of these, $k_v = 1$, has been used. Therefore the choice of any one parameter of the model, for instance, its geometric scale, completely determines the experimental conditions $\Big(k_w = \dfrac{1}{k_t}\Big)$. In addition, the transformation factor for the time is established $(k_t = k_l^2)$. However, we will start from the assumption that self-similarity occurs. In this case there are two degrees of freedom which are used as follows: it is assumed that $k_w = 1$ (identity of the kinematic conditions), and we will select the geometric scale of the model k_l. In contrast to the preceding, the time transformation factor will be $k_t = k_l$.

Now let us assume that the following experimental method is used. Pictures of the flow in the sample and model are taken. They are then brought to the same size by projecting the images onto screens, and synchronized by suitable changes in the speed of one of the films (homochronicity is converted into synchronization). It is obvious that under these conditions the two pictures should become identical. They can be superimposed by projecting them onto a single screen. Naturally, all this is true only to the extent to which self-similarity actually occurs. If the condition $\text{Re}'' = \text{Re}'$ is neglected, the modeling is only very approximate, and it is no longer possible to superimpose the pictures of the motion. It is impossible not to notice the differences, which become more and more marked as the degree of distortion increases.

34. THE METHOD OF ANALOGIES. THE GENERALIZED CASE AND THE PHYSICAL HOMOGENEITY OF PHENOMENA. PHYSICAL ANALOGIES. ANALOGOUS TRANSFORMATIONS

The model concept can be developed still further. Any two phenomena may exist in the relationship of original phenomenon and model if they are included in the same generalized case. This means that phenomena which are related as original and model must have the following properties in common (p. 34): after conversion to dimensionless quantities, the equations defining the phenomena and their boundary distributions must be identical, and the dimensionless parameters appearing in the equations as multipliers of the operators must have the same values. These features satisfy the requirements with respect to the common properties of the phenomenon and model, since they define unambiguously an individual generalized case. However, there is still another feature—the common nature of the physical natures of the phenomena. On isolating a special phenomenon from the generalized case, the groups are broken down into separate quantities. The various phenomena correspond to different methods of breaking down the groups into their components. The different variants are regarded purely as different numerical values of the factors, which in all cases must be regarded as quantities of the same physical types.

Thus it can be taken as standing to reason that a generalized case can include only phenomena which are of the same physical type. There is essentially no reasonable basis for this limitation. Of course, the idea of a generalized case arose in the search for a concept adequate for a group (collection) of phenomena of the same physical type (which are combined by the common nature of certain definite properties). However, in the form in which it exists ultimately, the concept of the generalized case, which is based on the consistent use of dimensionless quantities, is nevertheless quite foreign to the representation of the physical type of the phenomena. In a system of dimensionless quantities it is not even possible to consider the problem of their physical types. By means of the quantitative description a pure number is obtained which is not limited by a specific physical representation. It is only at the transition from the generalized case to the individual phenomena, when the dimensionless groups are broken up into separate dimensional quantities, that the problem of the physical natures of these quantities arises. At this moment, regardless of what has been said above, it comes about that each of the components is a quantity of the given physical type. The physical nature of the phenomena as a whole is determined in this way. It is obvious that this new condition represents an independent requirement in addition to the limitations by which the particular generalized case being considered is isolated from a whole mass of conceivable generalized cases. It is not related to the concept of the generalized case and does not result from its definition. In this sense it can be regarded as arbitrary. If the requirement of physical homogeneity of the phenomena included in the generalized case is excluded altogether, this can be expressed neither in the form of the generalized case, nor in the quantitative relationships which express the general nature of the properties of the individual concrete phenomena.

By giving up the requirement that the phenomena must be physically homogeneous, we can extend the limits of the individual generalized case. The range of phenomena which it includes is greatly extended, while the nature of the agreement between the phenomena is markedly changed. Any phenomena which are expressed identically in dimensionless form, regardless of their physical natures, are combined in the individual generalized case. This type of agreement between phenomena is termed a physical analogy. Obviously, physical analogy includes physical similarity within itself as a special case (phenomena of the same physical type). We can say that the individual generalized case (in its new, expanded definition) represents a group of phenomena which are all analogous to one another and that, within this group, they form smaller groups of phenomena which are similar to one another. The similarity relationships apply within each smaller group, while the analogy relationships apply between phenomena belonging to different groups. It is very important to note that these relationships differ only qualitatively. Quantitatively they are completely identical. Hence, if an investigation has a purely quantitative nature, any phenomena (within the given generalized case) can be compared as original and model. All the quantitative results obtained in the investigation of one can be directly transferred to the other. Thus the method of models develops into the method of analogies.

The method of analogies possesses obvious advantages. The use of the method of models (in its original form) allows us to change only the numerical values of the parameters of the process being studied, in particular, to increase or decrease the dimensions of the system, or to speed up or slow down the process. Even this is very convenient. However, changes in the very physical nature of the process being studied lead to vastly wider possibilities as regards both the setting up of the experiment and in the measurement methods. The method of analogies is widely and successfully used for investigating all manner of different problems.

Transfer phenomena in moving fluids give rise to a well-known example of analogies. The close relationship between the transport processes for heat (heat transfer), matter (mass transfer), and momentum (hydrodynamic resistance) is underlined by the name "triple analogy." It must be noted that in this case the analogy concept is narrowed considerably. All three of the effects are caused by the same process of displacing elements of the medium and, in the identity of their mathematical forms, their definitions reflect the real common nature of their physical mechanisms (which is not at all necessary).

A characteristic example of an analogy in the wider sense is the so-called electrothermal analogy, which in essence consists of replacing a steady-state temperature field which is to be studied by a steady-state electric potential field. In dimensionless form the equations defining the two fields are identical. The dimensionless boundary conditions are also identical, but only when they define directly the field of the unknown quantity at the boundaries of the system (i.e., when the thermal problem is set up with boundary conditions of the first or second type). It is impossible to derive an electrical analogy for boundary conditions of the third type. In this way, the group $Bi \equiv \frac{\alpha l}{\lambda}$ does not have an electrical analog. However, by representing it in the form $Bi = \frac{l}{\delta}$, we immediately overcome this difficulty, since it is obviously very easy to realize the requirement $\left(\frac{l}{\delta}\right)'' = \left(\frac{l}{\delta}\right)'$. Methods using the electrodynamic analogy are also widely used; this analogy makes it possible to study the properties of potential flow (i.e., the flow of fluids without friction) by means of an electrostatic model.

The electrical analogy is an exceptionally effective method of experimental investigation. As a rule, replacing the process to be studied by its electrical analog leads to appreciable advantages. In most cases it is easy to arrange an electrical model with given geometric and physical properties. The regime conditions required by the statement of the problem can usually be realized without particular difficulty even in the case of processes of a complicated nature. It is comparatively simple to carry out all the necessary measurements to a high degree of accuracy. It is easy to evaluate all the advantages of an electrical model as an object of experiment. Electrical models are particularly valuable in studying complicated unsteady-state processes (it is possible to achieve

electrical analogies for these only as a result of the large range
of the elements of electrical circuits and the flexibility of their
layout). As a very elementary illustration, let us look into the
analogy between the equation for a one-dimensional temperature
field:

$$\frac{\partial \vartheta}{\partial t} = a \frac{\partial^2 \vartheta}{\partial x^2}$$

and the equation for the distribution of the potential u in a conductor:

$$\frac{\partial u}{\partial t} = \frac{1}{RC} \frac{\partial^2 u}{\partial x^2},$$

where R is the electrical resistance and C is the capacitance.

Obviously the analog of the Fourier number $\frac{at_0}{l^2}$ is the group
$\frac{t_0}{RCl^2}$. It is easy to show how much the rate of development of the
thermal process differs from that of its electrical analog. Thus,
for example, we can find in the literature the results of investigations of annual variations in the soil temperature which have been
carried out by the electrical analog method. The temperature
variations immediately at the surface and at various depths (for
example, up to 10 m) are read from an oscillogram at intervals of
0.02 sec. In spite of the colossal speed-up of the process (the
time transformation factor is a quantity of the order of 10^{-9}), the
oscillograms give a clear picture of all its characteristic features.
In particular, we can see quite clearly the decrease in the amplitude
of the variations and the increase in the lag (phase displacement)
as we penetrate deeper into the soil.

The nature of the relationship between the concepts of physical
analogy and physical similarity are quite clear. In all cases the
main feature is the identity of the equations and the conditions for
single-valuedness, expressed in dimensionless form. If only this

feature is considered, we have defined an analogy. If it is combined with the requirement of physical homogeneity of the phenomena, the analogy is narrowed into the limits of similarity. Within the limits of similarity, the transition from a relative to an absolute form (from a generalized case to a specific phenomenon) is effected by multiplying each dimensionless quantity by parametric values (corresponding to the different phenomena), which are numerically different, but identical in physical nature. Correspondingly, the transition from one phenomenon to another leads to a similar transformation of all the quantities, i.e., leads to multiplying each of them by a transformation factor in the form of a pure number. Thus the transformation from one phenomenon to another is connected with a broader type of transformation than a similar transformation. It is customary to call this new operation (a similar transformation in the extended sense) an *analogous transformation.*

For a comparison of the fundamental process and its analog, let two corresponding quantities be y and z. In this case, we must have

$$\frac{y}{y_0} = \frac{z}{z_0}, \text{ so that } z = \frac{z_0}{y_0} y.$$

Thus the operation by means of which the quantity y is transformed into the quantity z consists of multiplying it (the quantity being transformed) by $\frac{z_0}{y_0}$. Let us introduce the notation

$$k_{zy} = \frac{z_0}{y_0}.$$

In the present case

$$z = k_{zy}\, y.$$

Without further explanation, it is clear that

$$k_{yz} = \frac{y_0}{z_0} \text{ and } y = k_{yz} z.$$

This type of multiplier (by means of which it is possible to transform not only the numerical value but also the physical nature of a quantity) is termed an *analogous transformation factor*. Obviously, an analogous transformation factor, in contrast to a similar transformation factor, is a dimensional quantity. The properties of the factors are identical in all other respects. Equations (4.6), which limit the freedom of choice of the numerical values of the transformation factors, remain valid for analogies also. However, in this case these equations have a deeper significance, since they interrelate not only the numerical values of the factors, but their dimensions also. The right-hand part of each of the equations is a dimensionless number (unity), and so the left-hand parts, which consist of groups formed from dimensional quantities, must also be dimensionless quantities. The actual significance of this condition, its importance, and its physical nature will be dealt with in more detail when we look into the method of investigation based on the analysis of dimensions.

Chapter V

DIMENSIONAL ANALYSIS

35. VARIOUS FORMS OF GENERALIZED ANALYSIS AND THEIR DEPENDENCE ON THE QUANTITY OF PRELIMINARY INFORMATION AVAILABLE

At the beginning of our discussion of general methods of setting up problems we noted that two possibilities arise in the process of quantitative analysis (p. 2-3). If a preliminary study of the phenomenon leads to a detailed model for it, and leads to the derivation of the fundamental equations, the subsequent quantitative investigation is carried out on these equations, which are regarded as being given by the statement of the problem. If we do not succeed in deriving the equations, the investigation must be based on relationships which are much less concrete. In all the various problems which have been studied, we have confined ourselves up to now to the first case, and have obtained quite a complete picture of the method of generalized variables and its application to processes represented by systems of fundamental equations. Now we must consider the way in which this method is applied when the fundamental equations cannot be set up.

Let us determine the main features of the problem facing us. The final purpose of the investigation has not changed. It is to develop a general method of setting up the variables which are characteristic of each process. The main difference in the new form of the problem is that now we no longer have the same dependable basis for determining the structures of the characteristic variables that was provided earlier through the equations expressing the most important quantitative properties of the processes. Under these conditions it is particularly important to establish the minimum amount of information which can still serve as the basis for setting up the problem of the construction of the characteristic variables. It is clear that we have to know exactly which quantities are important for the process, since the characteristic variables must be formed from these quantities. It is equally clear that, assuming we select only the important quantities (even if this is done completely and accurately), we will still not be in a state to pass on to the construction of the characteristic variables. The characteristic variables are essentially expressions of known relationships between the quantities from which they are made up. These relationships are caused by the mechanism of the process, and we must have some rational basis for their determination. Up to now we have found these bases in the equations which represent the mathematical model of the process. Now we are forced to limit ourselves to very general relationships of quite another type, and we will have to go into the derivation and significance of these more closely. To do this, let us return to a very early stage in the quantitative investigation, to the moment when the qualitative description of the process is being converted to a quantitative determination of its properties, and attempt to explain the basis on which this change is carried out.

36. THE TRANSITION FROM A QUALITATIVE DESCRIPTION
OF A PROCESS TO A QUANTITATIVE INVESTIGATION.
NUMERICAL VALUES OF THE QUANTITIES. THE OPERATION
OF COUPLING QUANTITIES WITH NUMBERS. PRIMARY AND
SECONDARY QUANTITIES

An investigation assumes a quantitative form when numbers
are introduced by means of the analysis, i.e., when the characteristics of the process become quantities represented by numbers. We can say that the quantitative principle is introduced into
the investigation with the numbers. It is therefore extremely
important to establish how the numerical values of quantities are
determined, i.e., the manner in which quantities are linked with
numbers. A consideration of this problem shows that there are
two different processes by means of which the characteristics of
a process can be reduced to numbers. Correspondingly, we have
to distinguish between two types of quantities.

Some quantities are introduced directly as characteristics of
the object of the investigation, without reference to any other
quantities. Length can be taken as a typical example of this sort
of quantity. The coupling of these quantities with numbers is carried out by means of an operation which can logically be called
direct measurement. Direct measurement consists in comparing
the quantity being measured with some accurately fixed quantity
of the same physical nature which has been chosen as a standard
and is known as the *unit of measurement.* The result of this comparison is a number expressing the relationship between the
measured quantity and the standard. Therefore, regardless of the
actual physical procedure (which is determined by the rules of
making measurements of the particular quantity), this operation

can be represented in the form $\dfrac{x}{x_0} = X$, where x is the quantity
being measured, x_0 is the standard value of the quantity, and X is

a number which is linked to the quantity x by means of the measuring operation. Conventionally we say that X is the numerical value of the quantity x, expressed in units of x_0. Obviously the numerical value of a quantity depends on the choice of the unit of measurement, and is inversely proportional to its size. Suppose that two different standards x_0'' and x_0' have been chosen for a quantity of the type x (i.e., suppose that we are using two different units), such that $x_0'' = c x_0'$. In this case, we obtain two different numerical values X' and X'' for each specimen of the quantity x; these are related by the obvious relationship $X'' = \dfrac{1}{c} X'$ or $X'' = kX$, where $k = \dfrac{1}{c}$.

Thus the numerical value of a quantity cannot be regarded as an absolute characteristic of the object of the investigation. This characteristic is relative in the sense that it changes depending on the choice of the unit of measurement. In contrast, the ratios of numbers corresponding to different concrete specimens of the quantity x do not depend on the choice of the unit of measurement. Actually, if we denote the different concrete values by subscripts, we have $X_1'' = kX_1', X_2'' = kX_2', X_3'' = kX_3', \ldots$, whence $X_1' : X_2' : X_3' \ldots = X_1'' : X_2'' : X_3'' : \ldots$. This very characteristic property of the operation of direct measurement can be described as the *absolute nature of ratios*.

It is easy to see that this property is of great importance, since the absoluteness of the ratios of the quantitative characteristics of real objects (i.e., their independence of the units of measurement and of the features of the measurement procedure used) can be taken as something which is quite indisputable and forms one of the physical concepts underlying any investigation. Thus it makes no difference at all what numbers are coupled as a result of measuring the height and diameter of a given cylinder, as long as these numbers are being considered separately; they may have quite different values depending on the choice of the units of measurement. However, if some definite unit of measurement is used in the measurements and if the ratio of the number corresponding to the height of the cylinder to the number corresponding to its diameter is five, this result expresses an objective property of the body being investigated. It cannot depend on the procedure by which it was obtained, and, of course, it cannot change depending on the choice of the units measured. Thus the absoluteness of the ratio is a necessary requirement which must be satisfied for any operation which is being used to link numbers to the quantities. In the case of direct measurements, this property is caused by the fact that changes in the unit of measurement necessarily lead to proportional changes of all the numerical values, where the proportionality constant is a quantity which is in inverse ratio to the units of measurement.

Quantities of this type possess a number of characteristic properties: 1) they are introduced irrespective of any other quantities; 2) they are coupled with numbers by the operation of direct measurement; 3) their units of measurement are chosen arbitrarily; 4) the condition of absoluteness of their ratios is satisfied automatically (as a direct result of the nature of the direct measurement operation). These are usually known as *primary quantities*.

From the definition of the concept of primary quantities, it follows that the operations for coupling primary quantities of various types with numbers cannot be interrelated in any form.

In addition to the primary quantities, other quantities with quite different properties are introduced. These are termed *secondary quantities*. The distinction between secondary and primary quantities is based on the radical difference in the operations by means of which the quantities are coupled with numbers. An indirect method is used for determining the numerical value of a secondary quantity; this is quite different from the method of direct measurement and usually much more complicated in concept. The numerical value of a secondary quantity is derived from the numerical values of several primary quantities according to rules which are established from the definition of the concept of this quantity. Velocity is a very simple example of a secondary quantity; it follows from the definition of velocity that its numerical value is obtained by dividing the number representing a length by the number representing time. Thus a correctly constructed definition of a secondary quantity should express sufficient information to settle the following problems: 1) the primary quantities to which the secondary quantity being considered is related; 2) what operations on the numbers corresponding to these primary quantities give the number representing the given secondary quantity. This information can be expressed clearly and concisely in the form of an equation defining the secondary quantity in terms of the appropriate primary quantities. This is usually referred to as the *determining equation*. Obviously, all determining equations are identical in nature.

Let us sum up. Each primary quantity can be defined completely without bringing in any quantities of other physical types. In this sense, all primary quantities are completely self-contained. In contrast, the definition of a secondary quantity takes on a real meaning only when it is associated with several primary quantities. From the very concept of a secondary quantity, the operation of coupling it with a number is reduced to definite operations on the numerical values of the primary quantities. This does not mean that the physical procedures which give the numerical value of the secondary quantity (the actual measurement of the secondary quantity) must necessarily reproduce the operations expressed in the determining equation. The main significance of the determining equation is that it expresses a strict logical system for forming the quantities. The quantities are introduced in stages in a definite succession: first independently (the primary quantities) and then on the basis of the appropriate relationships (the secondary quantities).

The collection of determining equations gives a complete representation of the procedures and principles for constructing the whole range of quantities which are of importance for the process. It is very important to note that this type of structural scheme cannot be regarded as something which is strictly fixed once and for all. The system of relationships underlying the formation of a quantity can vary depending on the physical content of the problem being investigated. Thus, if we are considering heat transfer in a constant volume of fluid moving at a moderate velocity, it is not possible to obtain any relationships between the quantities of the mechanical and thermal types. However, in the general case, when the problem is rendered more complex by the mutual transformation of heat and work, these two quantities (work and the quantity of heat) must be introduced by means of essentially identical determining equations. We can say that the main properties of the collection of important quantities—the number and type of the primary variables and the forms of the determining equations—depend on the specific features of the problem being investigated.

37. THE DETERMINING EQUATIONS. THE PRINCIPLE OF ABSOLUTENESS OF RATIOS AND THE STRUCTURE OF THE DETERMINING EQUATIONS

In spite of the great diversity of the determining equations, in some respects they are very closely related to one another. We can show very generally that the structure of any determining equation must satisfy a very stringent condition, the same in all cases. This condition arises in connection with the following obvious discrepancy. As we have explained, any operation used for coupling quantities with numbers must satisfy the condition of the absoluteness of the ratios. The requirement is also equally necessary for the direct measurement of the primary quantities and for the indirect methods which are used to obtain the numerical values of the secondary quantities. However, the situation is quite different in these two cases. If we are comparing numbers obtained as a result of the direct measurement of different specimens of a primary quantity, it follows that their ratio's independence are of the units of measurement is a direct and necessary consequence of the nature of the operation itself (since essentially a change to another unit of measurement results in *a proportional change in the numerical values)*. The property of absoluteness of ratios is a natural characteristic of direct measurement.

The conditions are quite different in comparing numerical values of secondary quantities. These values are obtained as the products of definite operations on the numbers representing several primary quantities. It follows directly from this that the numerical values of secondary quantities can vary, depending on the choice of the units of measurement for the corresponding primary quantities. It is necessary that the changes represent a *proportional transformation*: the requirement of absoluteness of ratios can only be satisfied under these conditions. However, the units of measurement of the various primary quantities can be chosen quite independently of each other.

It is easy to see that we are setting up a certain discrepancy here, since the freedom of choice of the units of measurement of the primary quantities is not generally speaking compatible with the predetermined nature of the change (a proportional transformation) of the numerical values of the secondary quantities. Obviously the resolution of this discrepancy must be sought in limiting the freedom of designating the operations which must be carried out on the numerical values of the primary quantities in determining the values of the secondary quantities. In other words, the freedom of choice of the units of measurement for the primary quantities can be retained only at the expense of limiting the structure of the determining equations. Let us express this limitation in analytical form.

Suppose that the determining equation is of the form

$$y = f(x_1, x_2, \ldots, x_m), \tag{5.1}$$

where y is some secondary quantity, and $x_1 x_2, \ldots, x_m$ are the primary quantities in terms of which it is expressed. The problem is to determine the form of the function f on the single condition that Eq. (5.1) must completely satisfy the requirement of absoluteness of ratios.

As we have seen, absoluteness of ratios exists automatically if the changes in y caused by transitions to other units of measurement are in the nature of proportional transformations. The specific meaning of this is as follows. According to Eq. (5.1), the numerical value of the secondary quantity (Y) is determined from the numerical values of the primary quantities $(X_1, X_2, \ldots, X_m)$ as the result of an operation represented in the form

$$Y = f(X_1, X_2, \ldots, X_m). \tag{5.1'}$$

If two different systems of units are being used:

$$x'_{10}, x'_{20}, \ldots, x'_{m0} \text{ and } x''_{10} = c_1 x'_{10};$$
$$x''_{20} = c_2 x'_{20}; \ldots; x''_{m0} = c_m x'_{m0},$$

we obtain two different sets of numerical values for the primary quantities:

$$X'_1, X'_2, \ldots, X'_m \text{ and } X''_1 = k_1 X'_1;$$
$$X''_2 = k_2 X'_2; \ldots; X''_m = k_m X'_m,$$

where $k_i = \dfrac{1}{c_i}$.

In the same way, two numerical values Y' and Y'' are obtained for the secondary quantity according to the equations

$$Y' = f(X_1', X_2', \ldots, X_m') \tag{5.2$'$}$$

and

$$Y'' = f(X_1'', X_2'', \ldots, X_m''). \tag{5.2$''$}$$

It follows from the condition that the transformation must be proportional that $Y'' = KY'$, where K is some still unknown but completely defined factor. In this case, the previous equations assume the form

$$\left. \begin{aligned} Y' &= f(X_1' X_2', \ldots, X_m'), \\ KY' &= f(k_1 X_1', k_2 X_2', \ldots, k_m X_m'). \end{aligned} \right\} \tag{5.3}$$

We have to find the form of the function f from these two equations when the factors $k_1, k_2, \ldots, k_m$ are chosen quite arbitrarily. This problem can easily be reduced to the other familiar problem, which we have treated in detail, of finding the structure of an absolutely homogeneous function.

If we digress from the real significance of the operations and compare only their numerical results, a proportional change caused by the transition to other units of measurement, and a similar transformation, corresponding to a transition to another similar phenomenon, can be regarded as completely identical operations. In both cases the numerical values change in accordance with $u'' = k_u u'$, where u' and u'' are the numerical values of some quantity u (it is unimportant whether this quantity is primary or secondary). If we concentrate on this relationship regardless of its derivation, it is quite unimportant why the numbers u'' and u' are related by the multiplier k: this may be due to the fact that the same quantity u is measured in terms of two different units, or that two different quantities u' and u'', forming similar groups, are measured in terms of the same units. Thus, quantitatively, the two operations are identical, though they are very different in content. Taking these differences into account, they can be distinguished as *metric transformations* and *physical transformations*.*

With this assumed nature of the agreement between metric and physical transformations, Eq. (5.3) assumes the significance of a system of relationships which define the function f as absolutely homogeneous. This means, however, that the number Y can be expressed in terms of the numbers X_1, X_2, ..., X_m only in the form of a power group

$$Y = A X_1^{a_1} X_2^{a_2} ... X_m^{a_m}, \tag{5.4}$$

where A is an arbitrary constant (p. 190).

We have found the only form of the relationship between the numerical values of primary and secondary quantities which satisfies the requirement of absoluteness of ratios. The structure of the determining equations is also solved. Obviously it must be

$$y = A x_1^{a_1} x_2^{a_2} .. x_m^{a_m}, \tag{5.5}$$

or

$$y = AD(x_1, x_2, ..., x_m), \tag{5.5'}$$

where D is a homogeneous operator defined by the relationship

$$D(x_1, x_2, ..., x_m) \rightarrow x_1^{a_1} x_2^{a_2} ... x_m^{a_m}.$$

Equations (5.5) and (5.5') are completely equivalent as expressions of the law by which quantities of the type y are constructed from quantities of the type x_1, x_2, ..., x_m. When represented in the form of a relationship between the transformation factors, they are identical in form:

$$K = k_1^{a_1} k_2^{a_2} ... k_m^{a_m}. \tag{5.6}$$

---

*A. A. Afanas'eva-Erenfest—one of the greatest experts on this problem—distinguishes these transformations as *formal* and *material*.

38. DIMENSIONS. THE FORMULA OF DIMENSIONALITY. DIMENSIONLESS QUANTITIES. THE INVARIANCE OF DIMENSIONLESS QUANTITIES WITH RESPECT TO METRIC TRANSFORMATIONS

In the derivation of Eqs. (5.4)-(5.6) we made no special assumptions about the properties of the quantities involved. The validity of these equations is limited in no way (apart from the necessary requirement that the ratios are absolute). As a result, they define some general properties of secondary quantities which are inherent in their natures as quantities defined in terms of primary quantities with fulfillment of the condition of absoluteness of ratios. It is remarkable that it is possible to derive equations which are so very general and which at the same time possess such a high degree of determinacy.

If another equation also contains the undefined constant A (we will consider this later), Eq. (5.6) expresses a strict single-valued relationship between the relative changes of the numerical values of the secondary quantities and the corresponding numerical values of the primary quantities on changing to another unit of measurement. We can see that only the exponents on the transformation factors of the primary quantities are of importance. These exponents are given by the determining equation. The exponents on the primary quantities are termed the dimensions of the secondary quantities with respect to the given primary quantities. The set of dimensions is recorded in the form of a *dimensionality formula*. Naturally, Eq. (5.6) can serve as a dimensionality formula. However, it has become customary to write this in another form—a symbolic equation—which is obtained from Eq. (5.6) by replacing the transformation factors of the quantities by their symbols (usually the symbol of a secondary quantity is enclosed in square brackets). For example, the dimensionality formula for velocity (symbol v) is written as

$$[v] = L T^{-1},$$

where L is the symbol for length and T is the symbol for time.

It is quite obvious that in the quantity and significance of the information which it contains, this formula is exactly equivalent to the equation written in terms of the transformation factors:

$$K_v = k_l/k_t.$$

Of course, it is quite valid to assume that the dimensionality formula contains the symbols of all the quantities which have to be regarded as primary from the physical content of the problem being investigated. In this case, the exponents on some of the primary quantities, particularly the exponents on the quantities which are not included in the operation used for determining the numerical value of a given secondary quantity, and which therefore do not appear in explicit form in the determining equation, must be put equal to zero.

The concept of dimensions can be extended (by convention, to some extent) to primary quantities, if it is assumed that the dimension of a primary quantity with respect to itself is unity, and with respect to any other primary quantity is zero. Thus, the dimensionality formula of a primary quantity always agrees with its symbol.

The indirect methods which are used for coupling secondary quantities with numbers and which underlie the definition of the concept of secondary quantities are capable of rational development. If a secondary quantity is defined in the necessary way, it can serve as a reference point for the formation of new secondary quantities. Thus a force is defined as the product of a mass and an acceleration (strictly, a force is a special type of physical quantity whose numerical value is defined as the product of the numerical value of a mass and the numerical value of an acceleration). Thus a force is introduced in agreement with the general principle of forming secondary quantities in terms of masses and accelerations. However, of these two quantities, only mass is a primary quantity in the normal type of dynamic problem, while acceleration is a secondary quantity, which in turn is defined in terms of length and time. However, this fact leads to no difficulty and is not a source of any confusion. Similarly, we do not experience any difficulty in going further in the same direction, and introducing quantities (work, impulse of a force) whose definitions must be based on the concept of force.

The concept of dimensionality retains its strict significance regardless of whether we are defining secondary quantities directly in terms of primary quantities or making use of other (previously defined) secondary quantities. This means that secondary quantities can appear as arguments only in the intermediate symbolic relationships. The final dimensionality formulas must be reduced to primary quantities. The corresponding transformation of the dimensionality formulas can be carried out much more simply, since the symbolic operations can be carried out accurately according to the same rules as the real operations on the numbers. This exact agreement between the symbolic and real operations is a direct result of the definition of the concept of dimensionality and obviously follows from the equivalence of the dimensionality formula and the equation for the transformation factors of the type (5.6). For force (symbol F), for example, we have

$$[F] = M[A] = MLT^{-2},$$

where M is the symbol for mass and $A \, (= LT^{-2})$ is the symbol for acceleration.

For work (symbol W) we find

$$[W] = [F]L = ML^2T^{-2}$$

and for the impulse of a force (symbol I):

$$[I] = [F]\, T = MLT^{-1}.$$

The case when all the exponents in the dimensionality formula become zero as a result of the symbolic operation is of special interest. In this case we say that the quantity being defined has *zero dimensions,* or is *dimensionless.*

Quantities of zero dimensions possess the important property of *invariance with respect to metric transformations,* (i.e., their numerical values remain unchanged when we pass to another unit of measurement). Actually, assuming that $a_1 = a_2 = \ldots = a_m = 0$ in Eq. (5.6), we find that $K = 1$, which is obviously evidence of the unalterable nature of the numerical value of the quantity y (for any values of k_i, i.e., for arbitrary changes in the units of measurement of the primary quantities). It is easy to see that the reverse situation is also true: if the numerical value of a quantity does not change on going to other units of measurement (of the primary quantities), it must have zero dimensions. Actually, the requirement $K = 1$ (constancy of the numerical value) is compatible with an arbitrary choice of the factors k_i only under the condition $a_1 = a_2 = \ldots = a_m = 0$. Thus, *all dimensionless quantities, and only dimensionless quantities, are invariant with respect to metric transformations.*

It is quite obvious that any ratio of two quantities of the same kind is a dimensionless quantity. However, it is not at all necessary that dimensionless quantities should be represented by this type of ratio. We are already quite familiar with examples of dimensionless quantities made up of quantities of different physical types, which are similarity groups and dimensionless variables of the type

$$\left(\frac{at}{l^2}, \quad \frac{\Delta p}{\rho w_0^2}, \quad \frac{al}{\lambda} \right)$$

Let us consider in greater detail the nature of the relationships which impart the property of nondimensionality to the complicated power groups made up of different types of quantities.

39. THE π THEOREM

Suppose that a group π is made up of primary quantities x_1, x_2, ... , x_m (total number m) and secondary quantities y_1, y_2, ... , y_r (total number r) in the form of a power expression:

$$\pi = x_1^{\alpha_1} x_2^{\alpha_2} \ldots x_m^{\alpha_m} y_1^{\beta_1} y_2^{\beta_2} \ldots y_r^{\beta_r} .$$

We are faced with the problem of expressing analytically the conditions necessary and sufficient for the group π to be a dimensionless quantity. The property of nondimensionality of π (i.e., its invariance with respect to metric transformations) can be represented in the form:

$$K_\pi = 1.$$

On the other hand, from the definition of the quantity π:

$$K_\pi = k_1^{\alpha_1} k_2^{\alpha_2} \ldots k_m^{\alpha_m} K_1^{\beta_1} K_2^{\beta_2} \ldots K_r^{\beta_r} .$$

Hence we must have

$$k_1^{\alpha_1} k_2^{\alpha_2} \ldots k_m^{\alpha_m} K_1^{\beta_1} K_2^{\beta_2} \ldots K_r^{\beta_r} = 1. \tag{5.7}$$

We still cannot draw any useful conclusions from this equation, since the factors K_j are not independent with respect to the factors k_i. We can only make the final decision after passing to the factors k_i, which can be chosen quite arbitrarily (i.e., after the left-hand part of the equation is reduced to a form equivalent to the dimensionality formula for the quantity π). This transformation can be carried out without difficulty. Each of the factors K_j is related to the factors k_j by an equation of the type (5.6):

$$K_1 = k_1^{a_{11}} k_2^{a_{21}} \ldots k_m^{a_{m1}} ;$$

$$K_2 = k_1^{a_{21}} k_2^{a_{22}} \dots k_m^{a_{2m}} ;$$

$$. \quad . \quad . \quad . \quad . \quad . \quad . \quad . \quad .$$
$$. \quad . \quad . \quad . \quad . \quad . \quad . \quad . \quad .$$

$$K_r = k_1^{a_{r1}} k_2^{a_{r2}} \dots k_m^{a_{rm}} ,$$

i.e., in general form,

$$K_j = k_1^{a_{j1}} k_2^{a_{j2}} \dots k_m^{a_{jm}} .$$
$$j = 1, 2, \dots, r .$$

Substituting these equations in Eq. (5.7), we find after elementary rearrangements that

$$k_1^{a_1 + a_{11}\beta_1 + a_{21}\beta_2 + \dots + a_{r1}\beta_r} k_2^{a_2 + a_{12}\beta_1 + a_{22}\beta_2 + \dots + a_{r2}\beta_r} \dots$$
$$\dots k_m^{a_m + a_{1m}\beta_1 + a_{2m}\beta_2 + \dots + a_{rm}\beta_r} = 1,$$

or, in more convenient form,

$$k_1^{a_1 + \sum_{j=1}^{r} a_{j1}\beta_j} \quad k_2^{a_2 + \sum_{j=1}^{r} a_{j2}\beta_j} \quad \dots \, k_m^{a_m + \sum_{j=1}^{r} a_{jm}\beta_j} = 1 .$$

This equation contains only transformation factors of the primary quantities which can by expressed independently (and hence the exponents represent the dimensions of the group π with respect to the corresponding primary quantities). Obviously the only way of satisfying it is to equate all the exponents to zero. Thus, we find:

$$
\left.
\begin{aligned}
&\alpha_1 + \sum_{j=1}^{r} a_{j1}\beta_j = 0, \\
&\alpha_2 + \sum_{j=1}^{r} a_{j2}\beta_j = 0, \\
&. \quad . \quad . \quad . \quad . \quad . \quad . \\
&. \quad . \quad . \quad . \quad . \quad . \quad . \\
&\alpha_m + \sum_{j=1}^{r} a_{jm}\beta_j = 0
\end{aligned}
\right\}
\qquad (5.8)
$$

or, in a more compact form,

$$\alpha_i + \sum_{j=1}^{r} a_{ji}\beta_j = 0.$$
$$i = 1,\ 2,\ \dots,\ m.$$

$$(5.8')$$

The system of equations (5.8) expresses the collection of conditions which are necessary and sufficient for the group π to be a dimenionless quantity. These conditions are expressed in the form of relationships between the exponents in the expression for π, thereby limiting the freedom of choice of their values. Thus, Eqs. (5.8) must be looked upon as a system of equations referring to the unknowns $\alpha_1,\ \alpha_2,\ \dots,\ \alpha_m;\ \beta_1,\ \beta_2,\ \dots,\ \beta_r$, in which the numbers $a_{11},\ a_{12},\ \dots,$ $a_{1m};\ a_{21},\ a_{22},\ \dots,\ a_{2m};\ \dots;\ a_{r1},\ a_{r2},\ \dots,\ a_{rm}$ (total number $r \cdot m$) are known directly from the dimensionality formulas for the quantities y, which play the part of constant coefficients given by the conditions. Any number satisfying these equations can serve as the exponent in the expression for the dimensionless group π. However, the total number of unknowns (n) is equal to the total number of primary and secondary quantities (i.e., $n = m + r$), while the number of equations is equal to the number of primary quantities only (i.e., m). The problem therefore has not one, but several solutions, the number of which is $n - m = r$.

This means that a collection consisting of n quantities of different types, of which m belong to the primary types and r to the secondary types, makes it possible to set up $n - m = r$ different dimensionless groups. This is a result of great importance, since the problem being investigated can be considered in a very general fundamental way.

It is assumed that we are studying a process about which we have the following preliminary information: 1) it is known how many quantities (number n) there are and which of them must be

regarded as important for the process; 2) the system of relationships in which the problem must be considered is known, and so
it is possible to establish the number (m) and type $(x_1, x_2, ..., x_m)$
of the primary quantities. From the result just obtained, it is
possible to determine directly the number of dimensionless groups
which are characteristic of the process in the form of the difference $n - m$. It is important to note that not all the primary
quantities in the necessary series must appear in the collection of
the n quantities which are important for the process. However, the
solution which is obtained remains valid in all cases, since it does
not depend on how many primary quantities there are and which of
them we do not place on the list of quantities which are of importance for the process.

Suppose, for instance, that the quantity x_k is not included among
the important quantities. In this case we have to assume $a_k = 0$ in
the corresponding equation, i.e., the equation for $i = k$. On doing
this, of course, this equation does not disappear, but only assumes
the simpler form

$$\sum_{j=1}^{r} a_{jk}\beta_j \equiv a_{1k}\beta_1 + a_{2k}\beta_2 + ... + a_{rk}\beta_r = 0.$$

Thus time, which is a primary quantity for an extremely wide
circle of problems, does not appear among the important quantities
in the case of a steady-state process. However, it is one of the
primary quantities in terms of which secondary quantities such as
velocity, power, thermal conductivity, and heat transfer coefficient
are defined. Thus the line in the system of equations (5.8) corresponding to time as one of the primary quantities is retained under
the conditions of a steady-state process.

We can see that under all circumstances the number of dimensionless power groups is equal to $n - m$. If all the primary quantities
appear in the list of quantities which are important for the process,

the difference $n - m$ coincides with the number of secondary quantities r which are important for the process, otherwise $n - m < r$. In all cases, we can state generally that *the number of dimensionless groups is equal to the number of all the quantities which are of importance for the process less the number of primary quantities*. This result is known as *Buckingham's π theorem*.

Here we have obtained the most general rule for determining the number of characteristic dimensionless groups for a process.

40. THE CONSTANTS OF THE DETERMINING EQUATIONS. FUNDAMENTAL AND DERIVED UNITS. DIMENSIONAL CONSTANTS. THE COMBINATION OF DIMENSIONALITY FORMULAS AND SYSTEMS OF UNITS OF MEASUREMENT

We have based our solution of the problem of the structure of the determining equations only on the principle of the absoluteness of ratios. No other arguments were presented which could serve as a basis for determining the nature of the relationships being investigated, but we were able to establish a strictly single-valued relationship between the metric transformation factors of the secondary quantities being formed and the primary factors. However, this quite clear result, which is presented in the form of Eq. (5.6), still does not give a complete solution of the form of the determining equation. As we have explained, these equations establish a special sort of relationship which allows us to draw concrete conclusions about only some (though very important) structural features of these equations. The incompleteness of the solution shows itself in the fact that the secondary quantities can be defined as a function of the primary quantities only to an accuracy of a constant factor, which remains quite arbitrary. This indefiniteness cannot be eliminated, since in the limit it is an idea which is related to the principle of the absoluteness of ratios. It is not difficult to show that on the basis of this principle alone it is impossible to go further in establishing a proportionality type of relationship. This relationship can be given the form of an equation by using proportionality coefficients which represent quite arbitrary constants. Additional arguments must be presented to determine these coefficients accurately; these arguments are actually related to the solution to the problem of the principles of construction of the units of measurement of secondary quantities.

We will discuss this as follows. Suppose that $x_1 = x_{10}$, $x_2 = x_{20}$, ... , $x_m = x_{m0}$, i.e., that in general, $x_i = x_{i0}$. In this case, $X_1 = X_2 = ... = X_m = 1$, and Eq. (5.4) gives us $Y = A$. Obviously, if the value of A is fixed, this will establish the principle of constructing the units of measurement of y. Of course, it would be more rational to take $A = 1$ as a universal convention. This convention is equivalent to the requirement that for all values of $x_i = x_{i0}$ we will also have $y = y_0$. Thus we have set up a general rule for constructing the units of measurement of secondary quantities based on the principle of simultaneous changes in the units of the primary quantity and of the secondary quantity being determined.

To sum up, we must note that, corresponding to the separation of quantities into primary and secondary, it is necessary to separate the units of measurement into fundamental (chosen arbitrarily) and derived units (formed from the latter according to definite rules).

Now the problem of the structure of the determining equations has been solved completely. However, can it be said definitely that the condition for the simultaneous change in the units of all the quantities (both primary and secondary defined quantities) is always satisfied? Obviously this condition can be satisfied in all cases when the determining equation is unique, i.e., when it contains secondary quantities whose units of measurement are known to be still not fixed. Under these circumstances, nothing prevents the corresponding choice of the units of measurement of the secondary quantities being defined.

The situation is quite different if there are two (or more) independent physical types of equations for some secondary quantity, and any of these can be chosen as the determining equation. This situation arises when the development of information leads to new, previously unknown, quantitative relationships which can be used as the basis for defining the secondary quantities introduced into the information earlier by means of other relationships. If it is possible to set up for some quantity two different equations of the type (5.4):

$$Y = A X_1^{a_1} X_2^{a_2} \ldots X_m^{a_m}$$

and

$$Y = B X_1^{b_1} X_2^{b_2} \ldots X_m^{b_m},$$

then any of them (but only one, of course) can be chosen as the determining equation. However, once the choice is made, the quantity y is determined in all respects: the dimensionality formula is established and the units of measurement are fixed. Suppose, for instance, that in one case the first equation is chosen as the determining equation. In this case the dimensionality formula for y is written in the form

$$[Y_*] = X_{1_*}^{a_1} X_{2_*}^{a_2} \ldots X_{m_*}^{a_m},$$

where the asterisk is used to indicate that we are considering not a numerical value, but a symbol of a quantity. The units of measurement of y must be defined in agreement with the general rule in such a way that the constant A tends to unity. At the same time it is obviously impossible for the constant B to tend to unity; after the units of measurement of y are fixed, B assumes a definite numerical value different from unity. It is easy to see that the numerical value of the constant B depends on the choice of the units of measurement of the primary quantities; i.e., it is by no means a dimensionless quantity. The dimensionality formula for B is obtained directly from the preceding arguments as a relationship of the form:

$$[B_*] = \frac{[Y_*]}{X_{1_*}^{b_1} X_{2_*}^{b_2} \ldots X_{m_*}^{b_m}},$$

or

$$[B_*] = X_{1_*}^{a_1 - b_1} X_{2_*}^{a_2 - b_2} \ldots X_{m_*}^{a_m - b_m}.$$

This type of quantity essentially represents a proportionality coefficient in the determining equation and cannot be excluded only because the units of measurement of the secondary quantities being defined have been chosen previously (on the basis of other equations). It is known as a *dimensional constant*. This name is a good indication of its main features—complete independence with respect to real changes of the physical conditions arising in the course of a process and, at the same time, variability due to changes to other units of measurement.

It is very interesting that it is possible in principle to eliminate the dimensional constant. Let us go back to the equation containing B. In contrast to what we did above, let us give it the significance of a relationship by which one of the primary quantities, x_m for instance, is defined in terms of the remaining primary quantities and the quantity y. This means that the quantity x_m is transferred into the category of secondary quantities, so that the relationship being considered becomes its determining equation. It is now possible to define the units of measurement of x_m in such a way that the constant B tends to unity. We see that the dimensional constant can be eliminated by transferring one of the primary quantities into the secondary category and correspondingly constructing a

derived unit of measurement for this quantity. In practice this
method is used in cases when it leads to no clash with previously
formulated systems of dimensions and units of measurement. Let
us consider a simple example to illustrate the ideas given here.

First let us consider the problem (already touched upon earlier)
of the nature of the correspondence between the quantity known as
the amount of heat and quantities of a mechanical nature. The
concept of a quantity of heat falls in the field of calorimetry, i.e.,
in a system of relationships which is quite unrelated to functions
of a mechanical nature. In investigating a process of heat redistribution in its pure form (transfer of heat in a solid body or in a
fluid stream of moderate velocity), when the initial calorimetric
representation of heat does not conflict with the physical content
of the problem, the quantity of heat should be included among the
primary quantities. But if there are effects involving the mutual
transformation of work and heat, we have to introduce the energetic
conception of heat with all the consequences attendant upon this.
In particular, the dilemma arises: either 1) the quantity of heat
must be relegated to the category of a secondary quantity, and in
this case the fundamental unit of measurement used for it (for instance, the calorie) must be replaced by a derived unit used for
work (for instance, the joule); or 2) the quantity of heat must remain among the primary quantities (retaining its initial unit of
measurement), and at the same time a dimensional constant (the
mechanical equivalent of heat) with the dimensions $ML^2T^{-2}Q^{-1}$ must
be included among the quantities which are of importance for the
process. Both the solutions are widely used in present-day practice, although the relegation of the quantity of heat to the category
of a secondary quantity (replacing calories by joules) leads to no
complications, so that the first solution is obviously preferable
(international agreement on this has already been obtained).

In the example just chosen, the elimination of the dimensionless constant causes no difficulties. A new physical law (the principle of equivalence of heat and work) establishes a relationship between the quantities belonging to two different systems of ideas (calorimetric and mechanical) which previously were completely independent of one another and had no point of contact. If the principle of equivalence is used as the basis of the determining equation for the amount of heat, this quantity passes from the primary category (in the system of calorimetric quantities) to the secondary category (in the system of mechanical quantities). No difficulty arises in doing this. In practice it reduces to the fact that a quantity of a mechanical nature—work—is introduced among the calorimetric quantities, replacing the quantity of heat in all the determining equations. Similarly, in constructing the derived units (for heat capacity, thermal conductivity, heat transfer coefficient, etc.) we introduce joules instead of calories, which raises more complicated problems for us.

In the system of primary units M, L, T, force is a secondary quantity, and is introduced by means of a determining equation based on Newton's second law. The resulting dimensionality formula has the form $[F] = ML T^{-2}$. However, force is also directly related to the primary quantities by the law of universal gravitation, according to which it is defined as a quantity proportional to the product of the masses of the interacting bodies and inversely proportional to the square of the distance between them. We have returned to the case of two different relationships, either of which can be used as the determining equation for the same quantity (the force f). Since the first of these is already taken as the determining equation, we retain a dimensional constant in the second, and write it in the form

$$f = \gamma \frac{m_1 m_2}{r^2},$$

where m_1 and m_2 are the masses of the interacting bodies, r is the distance between them, and γ is a dimensional constant, known as the *gravitational constant*.

The corresponding symbolic equation

$$[F] = [\Gamma] M^2 L^{-2}$$

leads directly to the dimensionality formula for the gravitational constant:

$$[\Gamma] = [F] M^{-2} L^2 = M^{-1} L^3 T^{-2}.$$

If we wished to get rid of the gravitational constant, this would lead to the necessity of relegating one of the primary quantities to the secondary category (for instance, it could be assumed that $[M] = L^3 T^{-2}$). It is easy to see that in this case there will inevitably be a clash of all the systems of dimensions (and, correspondingly, of the units of measurement). In contrast to the previous example, therefore, we have to admit that it is definitely unsuitable to eliminate the dimensional constant.

The numerical value of the gravitational constant naturally depends on the choice of the main units of measurement. In the CGS (cm, g, sec) system:

$$\gamma = 6.87 \cdot 10^{-8} \ \frac{cm^3}{g \cdot sec^2} \cdot$$

We have arrived at a definite system for representing the properties of the quantities which are introduced as quantitative characteristics of a process, and the nature of the relationships which interrelate these quantities in the most general case. All the quantities which are of importance for a process can be divided into two groups: the first consists of variables and the second of dimensional constants. The number and nature of the variables are determined exclusively by the physical model of the process which is assumed in the investigation. The number and nature of the dimensional constants also depends on the system of dimensions in terms of which the problem is considered. In this sense, the variables and dimensional constants can be distinguished as physical and metric quantities.

The system of dimensions is entirely determined by which variables are selected as primary quantities. The classification of quantities (their division into primary and secondary) depends to a great extent on the content of the problem. However, to some extent a freedom of choice remains in the matter. The transition from a set of dimensionality formulas to a system of units of measurement is based on the principle of simultaneous changes in the units of the secondary quantities being defined and of the primary quantities appearing in the determining equations. Any divergence from this principle leads to the appearance of dimensional constants. The dimensionality formulas provide the basis for determining the structures of the units of measurement of the secondary quantities (i.e., the rules for forming the derived units from the fundamental units), but not their dimensions. Essentially, no limitations can be made, because of the initial idea of the freedom of choice of the dimensions of the primary quantities (and hence of the derived quantities also). Thus, an infinite number of systems of units of measurement are possible within the limits of a given system of dimensions which are identical in structure, but different in dimensions. These systems also differ in the numerical values of the dimensional constants.

The dimensionality formula is an effective means of investigating problems of a metric nature. It can be used successfully for calculating the numerical values of quantities when the dimensions of the fundamental units are changed (for example, on going from a cm, g, sec system to a m, kg, hr system) or in more complicated cases—on going to another system of primary quantities (for example, from $ML\bar{T}$ to FLT).* A consideration of these applications would take us too far from our main objective. We will concentrate on the problem of constructing generalized variables by using the apparatus of dimensional analysis.

41. THE APPLICATION OF THE APPARATUS OF DIMENSIONAL ANALYSIS TO THE CONSTRUCTION OF GENERALIZED VARIABLES. THE DIMENSIONALITY FORMULA AS A FORM OF REPRESENTING THE REDUCED GROUPS FOR THE DETERMINING EQUATIONS

We will argue as follows. If we knew the equations defining the process we could derive the generalized variables with no difficulty. However, with a problem stated in the present way, we have to assume that we are given only a list of the important quantities and their dimensions. Let us show that this information is sufficient for inferring the structure of the unknown fundamental equations of the problem to the extent that is necessary for determining the form of the generalized variables. To do this we must first note that all equations which represent a mathematical model of a process which is adequate for its physical mechanism have the following quite general and very important property: when the dimensions of the fundamental units are changed arbitrarily, the form of the equation does not change. Equations possessing this feature are usually said to be *complete*. Obviously the property of completeness means that the equations cannot contain coefficients of the metric type, i.e., quantities which do not characterize any physical aspect of the process. In this respect the only exceptions (which do not prevent completeness of the equations) are dimensional

---

*The use of the dimensionality formula for calculating purposes is limited to cases of systems of primary quantities which are uniform in number (but not in nature). The equality in number of the group of primary quantities (the coherence of the system) is a condition sufficient for the existence and determinacy of a solution. In the cases of systems not satisfying this condition, the solution, generally speaking, is either indeterminate or does not exist.

constants, which are introduced through the determining equations (when it is impossible to eliminate them) and which are included from the start among the quantities which are of importance for the process. Thus we start from the assumption that our unknown equations are complete in all cases. But in this case they must have a structure which gives them invariance with respect to similar (metric) transformations.

We have arrived back among familiar ideas, and the further course of the discussion is quite clear. The equations can have the property of invariance with respect to similar transformations in two cases only: 1) the reduced groups corresponding to the separate terms of the equations have the same dimensions and are transformed into a dimensionless power expression by combining them in pairs in the form of ratios (this is the main case which we have been dealing with all along), or 2) the reduced groups are obtained directly as quantities of zero dimensions. We therefore maintain that there are only two possible assumptions in connection with the fundamental equations of the problem: either the left-hand part represents the sum of terms of the same dimensions (i.e., the equation is dimensionally homogeneous), or it consists of an expression of zero dimensions. In both cases the problem being considered (the construction of the generalized variables) reduces to the determination of the dimensionless groups which can be formed from the given group of quantities which are of importance for the process (i.e., it reduces to a problem which has been analyzed already).

The number of dimensionless groups is obtained directly on the basis of Buckingham's π theorem. In order to determine the forms of the groups, we have to solve a system of algebraic linear equations (the total number of these corresponds to the number of primary variables) with a large number of unknowns;

i.e., it is necessary to carry out a fair amount of tedious computational work. However, it is possible to use another route which is technically more simple and economical.

The dimensionality formulas and the equations for the transformation factors corresponding to them are completely equivalent to one another, as we have already shown repeatedly. If the dimensionality formula of some quantity y is

$$[Y_*] = X_{1_*}^{a_1} X_{2_*}^{a_2} \ldots X_{m_*}^{a_m},$$

the equation for the transformation factors is written in the form

$$K = k_1^{a_1} k_2^{a_2} \ldots k_m^{a_m}, \text{ or } \frac{K}{k_1^{a_1} k_2^{a_2} \ldots k_m^{a_m}} = 1.$$

In turn, this corresponds to the reduced group

$$\pi = \frac{y_0}{x_{10}^{a_1} x_{20}^{a_2} \ldots x_{m0}^{a_m}}.$$

Thus, *the dimensionality formula is a special form of writing the reduced group corresponding to the determining equation.* Thus, for instance, the dimensionality formula for velocity $[V] = L T^{-1}$ is written in the other representation as the well-known homochronicity group $\frac{v_0 t_0}{l}$. Similarly, the dimensionality formula for force $[F] = M[A]$ corresponds to the group $\frac{f_0}{m_0}$, and so in connection with $[A] = L T^{-2}$ $\left(\text{and correspondingly, } \frac{a_0 t_0^2}{l}\right)$, we arrive directly at the group

$$\frac{f}{ma_0} \cdot \frac{a_0 t_0^2}{l} = \frac{f_0 t_0^2}{ml}.$$

It is interesting to turn our attention to the following relationship, which is obtained by multiplying the numerator and denominator of this group by v_0:

$$\frac{f_0 t_0^2}{ml} = \frac{v_0 t_0}{l} \cdot \frac{f_0 t_0}{m v_0}.$$

In this way the similarity group corresponding to the determining equation for a force can be represented as the product of the homochronicity group and a group corresponding to the equation for the impulse of a force.

Bearing in mind the relationship between the dimensionality formulas and the reduced groups, we can obviously use the following procedure for constructing the generalized variables. The reduced groups are set up directly from the dimensionality formulas. If some of the primary quantities do not appear among the quantities which are of importance for the process, these can be excluded. As a result, we obtain a complete system of dimensionless groups. Let us show that this assumed procedure actually leads to a collection of groups whose number is in agreement with the π theorem.

Suppose that the total number of important quantities is n, of which r are secondary quantities. The problem is being considered in terms of a system of dimensions with the number of primary quantities equal to m. The number of determining equations, and hence of dimensionless groups, which can be set up is obviously equal to r. If all the primary quantities are important in the process (i.e., $n = r + m$), all the groups obtained are included in the group of unknown generalized variables. In this case, their number is determined directly from the difference $n - m = r$. If some of the primary quantities are unimportant, the number n will decrease for a given value of r. However, at the same time the number of groups decreases, since the unimportant primary quantities can be eliminated, and each elimination operation reduces the number of groups by one. The final number of groups will therefore be determined as before in the form of the difference $n - m$ (but this will no longer be equal to the number of primary quantities).

For example, suppose that two primary quantities are unimportant in a process. In this case, $n = r + m - 2$. In all, r dimensionless groups will be obtained. When the unimportant primary quantities are eliminated the number of groups decreases by two. As a result, the required collection of generalized variables will consist of $r - 2 = n - m$ terms. We can see that Buckingham's π theorem is satisfied in all cases.

By way of illustration, let us consider the well-known problem of hydraulic resistance in the steady-state flow of an incompressible fluid along a channel. The unknown quantity is the pressure drop Δp. Apart from this, the quantities of importance for the process are: the length l, the velocity w, the density ρ, and the viscosity μ. Apart from the unknown quantity (Δp), all these quantities are represented by parametric values in the conditions. The problem is typically mechanical in its content. We will therefore consider it in terms of the MLT system of primary quantities. It is unnecessary to introduce any dimensional constants. In this way, we have $n = 5$ for the total number of quantities which are important in the process. The number of primary quantities is $m = 3$. The number of dimensionless groups characterizing the process will be $n - m = 2$.

We can make up the following table:

Quantity	Dimensionality formula	Reduced group
Length l	Primary quantity	
Pressure drop Δp	$[p] = ML^{-1}T^{-2}$	$\dfrac{\Delta p l t_0^2}{m} \equiv \pi_1$
Velocity w	$[w] = LT^{-1}$	$\dfrac{w_0 t_0}{l} \equiv \pi_2$
Density ρ	$[\rho] = ML^{-3}$	$\dfrac{\rho l^3}{m} \equiv \pi_3$
Viscosity μ	$[\mu] = ML^{-1}T^{-1}$	$\dfrac{\mu l t_0}{m} \equiv \pi_4$

The primary quantities m and i do not appear among the important quantities, and can be eliminated. This can be done, for instance, in the following way:

$$\frac{\pi_1}{\pi_3\pi_2^2} = \frac{\Delta p}{\rho w_0^2} \equiv \text{Eu},$$

$$\frac{\pi_2\pi_3}{\pi_4} = \frac{\rho w_0 l}{\mu} \equiv \text{Re}.$$

We have therefore arrived at known expressions.

42. THE RELATIONSHIP BETWEEN THE THEORY OF SIMILARITY AND DIMENSIONAL ANALYSIS

Thus by using the apparatus of dimensional analysis the problem of the structure of the generalized variables is solved as follows. The type of the problem is determined, and the system of dimensions is chosen. A list is made up of the quantities which are of importance for the process, including dimensional constants. The number of generalized variables is determined (as the difference between the total number of quantities and the number of primary quantities). The dimensionality formulas are transformed into the power groups. The primary quantities which do not appear in the list of quantities important for the process are eliminated. We have now reached the end of the steps in the solution which are connected with the application of the ideas of the theory of dimensions. All the remaining operations are based on considerations which are entirely unrelated to the concept of dimensionality.

It is easy to see that two aspects of this scheme are of primary importance: 1) the determination of the type of problem (and the corresponding choice of the system of dimensions) and 2) the preparation of the list of the important quantities. The quality of the solution as a whole is related to how well we succeed in solving these two problems. Everything else reduces to elementary technical operations. If the collection of primary quantities is determined incorrectly because of some invalid assumption of the physical content of the problem (if some quantity is wrongly regarded as being a primary quantity or, conversely, if some essentially primary quantity is not included among the primary quantities), this will greatly impair the quality of the solution. In the first case, we will obtain an incomplete system of dimensionless groups (since in determining the difference $n - m$ the quantity calculated will be incorrectly increased by unity) and so one of the arguments will disappear from the generalized equation. This means that the most important property of the equation will be lost—the single-valuedness of the relations expressed by it, and the solution which is obtained is of no practical value. In the second case, on the other hand, we get an extra dimensionless group. Correspondingly, there is an additional argument in the generalized equation, and the solution is unjustifiably weakened. The errors made in determining the number and types of quantities important in the process are no less troublesome. These difficulties can be avoided only by a sufficiently detailed knowledge of the physical mechanism of the process being studied. To state the problem correctly, therefore, it is necessary to have available the corresponding physical representation.

The problem of the quantity of preliminary information which is necessary is of the greatest interest to us, since the difference between the two forms of generalized analysis—the theory of similarity and dimensional analysis—is mainly one of the amount of preliminary information available. The applicability of the theory of similarity depends largely on the possibility of setting up the problem correctly (in the analytical sense) or, at least, of setting up the system of fundamental equations and formulating the conditions for uniqueness of solution from physical considerations.

The prerequisites for the applicability of dimensional analysis are much less rigid. The important feature of dimensional analysis as a form of generalized analysis is that it is based on an apparatus which does not need the equations of the problem to be included. To apply the method of dimensional analysis, it is sufficient to know the system in which the quantities which are of importance for the process fall. The equations of the process represent an exact quantitative expression of the assumed physical model of the process being investigated. The treatment of these equations using the apparatus of the theory of similarity makes it possible to derive generalized variables in the form of groups of the quantities which are combined into a whole by means of relationships based on the model of the process and expressed in implicit form in the equations. This direct path is completely closed in the case of dimensional analysis, which does not depend on the possibility of constructing a sufficiently detailed model of the process for setting up equations. The apparatus of dimensional analysis disposes of means for determining the generalized variables on the basis of relationships of a much more general nature in the form of the determining equations (or dimensionality formulas).

Thus, the essential difference between the theory of similarity and dimensional analysis is that *the apparatus of the theory of similarity is applied to the equations of the process, while the apparatus of dimensional analysis is applied to the determining equations (dimensionality formulas).*

If we bear in mind the formal techniques and fundamental content of the methods, we can say that essentially the same apparatus is applied with different objectives. It is therefore somewhat different in technical respects.

We can see that the dividing line between the theory of similarity and dimensional analysis is the difference in the completeness of the physical information that is required about the process. Generalized analysis in any form must be based on a clear representation of the mechanism of the process. However, in using the theory of similarity a greater amount of preliminary information is required, since in this case the physical ideas must reach a depth and concreteness sufficient for deriving equations which define the process. If the theory of similarity can be applied, it should certainly be given preference in view of the fact that its apparatus is simpler. In addition, the theory of similarity also includes an explanation of the significance of the dimensionless groups involved.

If the equations of the problem are not known we are forced to use dimensional analysis. Under these conditions we are not always completely confident that we can make up a correct list of the quantities of importance for the process or make a valid choice of the system of dimensions. However, when dimensional analysis is based on a correct choice of the group of important quantities and when the system of determining equations is correctly set up, it becomes an investigational tool by no means inferior to the theory of similarity with respect to the accuracy, completeness, and concreteness of the results it produces.

Dimensional analysis is based on a consideration of relationships of a very general nature and rests on concepts which are notable for their high degree of abstraction. The results obtained by using it express profound physical ideas and to a large extent are universal in nature. Some of these results—the proof of the power structure and of the zero dimensions of the groups, and the establishment of a general method for determining the number of groups (the π theorem)—throw light on the fundamental aspects of generalized analysis. Since they can be proved in such a general way, they also play an important part outside the strict limits of dimensional analysis.

SUBJECT INDEX

255